6.00

CLINICAL
DIAGNOSTIC
PEARLS

by

JASON C. BIRNHOLZ, M.D.

Department of Radiology
Massachusetts General Hospital
Boston, Massachusetts

with

PAUL E. MICHELSON, M.D.

Wilmer Ophthalmologic Clinic
Johns Hopkins Hospital
Baltimore, Maryland

MEDICAL EXAMINATION PUBLISHING COMPANY, INC.
65-36 Fresh Meadow Lane
Flushing, New York 11365

Library of Congress
Catalog Card Number
75-146990

ISBN 0-87488-730-5

October 1971

PREFACE

> Given the sacred hunger and the proper preliminary training, the student practitioner requires at least three things with which to stimulate and maintain his education: a note-book, a library, and a quinquennial brain dusting.   (William Osler)

It is the authors' intention that this collection of notes serve as a nucleus about which the medical student may build his own treasured 'book of pearls'. The peculiarities and salient features of human disease are often elusive memories for the neophyte until his knowledge and judgment are tempered by valuable clinical experience. In the interim, short, concise notes are convenient memory clues and a useful educational adjunct.

The material is arranged into categories - 'presenting symptoms and regional physical findings', specific diseases, laboratory findings, and drug effects. The authors wish to explain in advance that the selection (and arrangement) of material is arbitrary and reflects their own particular interests. There has been no attempt at text-book detail or encyclopedic thoroughness. A substantial background of factual knowledge has been assumed; emphasis has been directed toward some of the curious and less well-known aspects of medicine which are of practical interest and of clinical value.

The reader is urged to examine each statement with the critical framework of his own knowledge of pathophysiology. Wherever possible, reference is made to the individual or article associated with a specific fact; apologies are offered for that material which has become 'common knowledge' and for which the primary author is not credited. When looking up specific material, the reader should consult the index, since any one topic may appear under several headings.

Finally, please recall that as information is condensed there is necessarily some sacrifice of accuracy. There are always exceptions, modifications, or explanations that an outline such as this must omit. This collection of notes has been a pleasure rather than a task to compile; hopefully, it will be a pleasure for the reader as well.

The authors wish to thank Dr. Victor A. McKusick and Dr. David Paton of the Departments of Medicine and Ophthalmology of the Johns Hopkins University School of Medicine for their guidance.

This book is dedicated to

our parents

and to Elaine Elizabeth and Shelly

Errors like straws upon the surface flow,
He who would search for pearls must dive
below.

Dryden

TABLE OF CONTENTS

FOREWORD

DIVERSE PREFATORY THOUGHTS

Subjects arranged alphabetically
within each section

FOREWORD

Several years ago, while medical students here at Johns Hopkins, Birnholz and Michelson began collecting "pearls" from their clinical teaching. This collecting has continued as they progressed through the successive steps of their training -- Birnholz in internal medicine, Michelson in ophthalmology. The digested results are served up here for the edification of all of us who are in training -- no "compleat physician" ever really completes his period of training.

The product of the Birnholz-Michelson pearl-collecting is rather like Joseph H. Pratt's A Year With Osler, except that the latter did not see the light of day until about 50 years after Pratt's exposure to the master as a medical student. It is something like A Year in Osler (meaning a year in the Osler Clinic) which I got together for private and limited circulation, to review the clinical experience by the Ward Medical Service at Johns Hopkins during the year 1951-1952 that I was the resident.

It is also a bit like the Aphorism of Hippocrates -- which were really "clinical pearls" of a school of clinicians -- and the writing of Peter Mere Lathan and Armand Trousseau, great clinical teachers in London and Paris, respectively, in the last century. And of course they are like the axioms extracted from the writing of Osler by his devotee William Bennett Bean.

An aphoristic approach to medicine may have been useful 500 years before Christ, but is there a place for this approach in this day of computerization of diagnosis? I say emphatically yes. The computer is powerless if there is nothing to go into it. Clinical pearls serve a useful function, if a hardnosed attitude is taken toward them, i.e., the truth of the statement continually questioned and its quantitative significance sought. For example, heliotrope coloration of the skin about the eyes is a sign of dermatomyositis. The question is not only is this true, but also in what proportion of cases of dermatomyositis is it true, and in what other conditions is it also found thus leaving room for confusion? Catchy clinical pearls must not be accepted as dogma and must not be permitted to encourage a superficial, "once-over-lightly" approach to clinical problems in lieu of a painstaking and studious analysis.

The authors have obviously had a lot of fun in assembling this material, and you the readers will, I am sure, have the same pleasure I have enjoyed when you go through it and use it as a clinical vade mecum.

Victor A. McKusick,
Johns Hopkins Hospital
Baltimore, Md. 21205

DIVERSE PREFATORY THOUGHTS

Hippocrates:

Life is short and the art long; the occasion fleeting; experience fallacious, and judgment difficult. The physician must not only be prepared to do what is right himself, but also to make the patient, the attendants, and externals cooperate.

Where there is love of humanity, there is also love of the art of medicine.

Celsus:

The art of medicine has no constant rule.

Sydenham:

The physician should bear in mind that he himself is not exempt from the common lot but subject to the same laws of mortality and disease as others, and he will care for the sick with more diligence and tenderness if he remembers that he himself is their fellow sufferer.

Bernard:

When we begin to base our opinions upon medical fact, on inspiration, or on more or less vague intuition about things, we are outside of science and are exemplars of that fanciful method fraught with the greatest dangers in that the health and life of the patient turn upon the whims of an inspired ignoramus. True science teaches us to doubt and, in ignorance, to refrain.

Osler:

Even in populous districts the practice of medicine is a lonely road which winds uphill all the way, and a man may easily go astray and never reach the Delectable Mountains unless he early finds those shepherd guides of whom Bunyan tells, Knowledge, Experience, Watchful, and Sincere.

SYMPTOMS AND REGIONAL PHYSICAL FINDINGS

Hutchison (B M J 1:335 '28):

There is, I believe, a real danger lest the increased use of laboratory tests, X-ray examinations, and other short cuts to diagnosis should lead to neglect of the information to be obtained by the skillful use of the unaided senses, and to a comparative atrophy of these from disuse.

Longcope (Bull JHH 50:4 '32):

Pope has said that the "proper study of mankind is man" and even though a clinician has science, art, and craftsmanship, unless he is intensely interested in human beings he is not likely to be a good doctor.

ALOPECIA

Alopecia of the outer third of the eyebrows (St. Anne's sign) is seen with syphilis, hypothyroidism, and leprosy. With leprosy, eventually, complete facial alopecia occurs. Sparse eyebrows occur with hypoparathyroidism.

Spotty alopecia of the temporal zone is a feature of secondary syphilis; frontal and temporal loss occurs in sickle cell anemia; and occipital alopecia is seen with tuberculosis. Spotty scalp alopecia without predilection for region occurs with collagen vascular disorders and with hyperthyroidism.

Total scalp and eyelash alopecia follows chronic arsenic or thallium poisoning and is seen in association with the multiple endocrine adenoma syndrome and with Werner's Syndrome.

Post-febrile hair loss was formerly most often seen after typhoid fever; it may occur after any septic condition with high, spiking fevers. The hair buds are not damaged, and complete regrowth will take place.

Scalp hair thinning to complete alopecia may follow one to three months after heparin use. The mechanism appears to be similar to that of post-febrile hair loss.

A narrow longitudinal band of hair loss ascending from the middle of the forehead in a young person (coup de sabre alopecia) is characteristic of scleroderma.

ANGIOID STREAKS (Retina)

Angioid streaks (cracks in Bruch's membrane, not to be confused with choroidal rupture after, for example, orbital trauma) are found (with the hand ophthalmoscope) in more than 85% of cases of pseudoxanthoma elasticum. They are also found in some 15% of cases of advanced Paget's disease of bone and are seen rarely with:
1. sickle cell anemia
2. lead poisoning
3. thrombotic thrombocytopenic purpura
4. Ehlers-Danlos syndrome
5. senile elastosis

ARRHYTHMIAS

Polyuria occurs with paroxysmal tachycardias, particularly paroxysmal atrial tachycardia. When an history of polyuria following palpitations is obtained, it is likely that this arrhythmia has occurred.

Four or more heart sounds are often distinctly audible in persons with ventricular tachycardia and represent a combination of gallops and split first and second sounds.

The electrocardiogram will not always distinguish ventricular tachycardia from supraventricular tachycardia with aberrant conduction. In the former state, however, atrio-ventricular dissociation engenders a number of valuable physical signs:
1. variation in the intensity of the first sound during held inspiration.
2. irregular cannon A waves in the jugular venous pulse.
3. irregular intensity of Korotkoff sounds at fixed sphygmomanometer pressure.

THE CHEST

Bilateral anterior prominence of the upper two thirds and indrawing of the lower third of the chest wall (the thorax of Davies) is said to occur in some 50% of cases of ventricular septal defect and 10% of cases of Fallot's tetralogy (Am J Cardiol 20:309 '67). A unilateral parasternal bulge in the region of the 2nd rib indicates an atrial septal defect, in the region of the 5th and 6th ribs, a ventricular septal defect (ibid).

Lateral displacement of the nipples is associated with bilateral renal hypoplasia (J Ped 69:806 '66) and with Turner's syndrome. Accessory nipples are found in about 15% of patients with congenital cardiac anomalies (JAMA 158:821 '55).

Scoliosis appears to occur in association with congenital heart disease, being present in as much as 6% of cyanotic disorders (less often with acyanotic defects) (J Ped 73:725 '68).

Scratch resonance is a useful technique in confirming the presence of pneumothorax (NEJM 264:88 '61).

Continuous bruits over the chest may be due to:
1. venous hum
2. patent ductus
3. coronary sinus fistula
4. coronary, subclavian, or mammary A-V fistula

The venous hum is usually noted in the region of the right sternoclavicular joint. It is loudest with the patient sitting up, and it is abolished with maneuvers which decrease jugular venous flow (such as compression). Although venous hums of the chest occur in "normals," such a finding should prompt a consideration of situations with increased flow, namely, anemia, large A-V fistula, or hyperthyroidism.

CLUBBING (see also HYPERTROPHIC OSTEOARTHROPATHY and PACHYDERMOPERIOSTOSIS)

In addition to intrathoracic and cardiovascular causes, about 10% of all cases of bilateral clubbing are secondary to:
1. chronic gastrointestinal disease, particularly ulcerative colitis, regional enteritis, biliary cirrhosis, or cystic fibrosis.
2. gastric, esophageal, or renal neoplasms
3. chronic myelogenous leukemia
4. ascariasis, malaria
5. some toxic metallic or metalloid agents, such as arsenic, phosphorus, mercury, and beryllium.

Clubbing is uncommon in uncomplicated pulmonary tuberculosis, but occurs in as much as 25% of cases with supervening chronic lung disease.

Painful clubbing is usually secondary to a malignancy. In a Mayo Clinic series, painful clubbing was seen with 50% of pleural tumors, 10% of primary lung cancers, and 3% of metastatic lung tumors (Dis Chest 52:62 '67).

Clubbing can occur with mitral stenosis and is an important physical sign of bacterial endocarditis.

Asymmetric clubbing indicates impaired regional vascular supply. With clubbing of the toes but not of the hands, coarctation is suggested. With clubbing of one hand, brachial occlusion or diversion of flow (as with an A-V fistula or the subclavial steal syndrome) are possibilities. Moreover, with clubbing of one hand, Pancoast's tumor must be excluded.

Unidigital clubbing is seen with sarcoidosis.

Recurrent clubbing may sometimes be seen during pregnancy in women who are otherwise healthy. Clubbing may, of course, be familial (congenital), in which instance it is of no clinical import.

COMA

Immediately after drawing blood for the laboratory, the undiagnosed individual should be given 25 grams of dextrose intravenously as the 50% solution. This amount of sugar is not harmful if there is hyperglycemia, but it may well be lifesaving if there is hypoglycemia.

Consider lactic acidosis when the comatose individual is in acidosis with a wide anion gap (i. e. (Cl + HCO3   Na + K - 16)). Lactic acidosis occurs in diabetics and may accompany both ketoacidosis and hyperosmolar coma. It is also seen with chronic pancreatitis, uremic coma, and ethylene glycol intoxication. Dialysis is the therapy of choice, followed by intravenous bicarbonate.

Tear glucose increases as serum glucose increases, with values above 10 mg% indicating serum glucose above 150 mg%. Screening the comatose patient for glycolacria with Clinstix (Am J Ophth 65:414 '68) is a very rapid means of detecting hyperglycemia.

Nonketotic hyperosmolar coma may be the initial presentation of an elderly, previously undiagnosed diabetic. Serum glucose may even surpass 1000 mg%. Serum osmolarity may be estimated by the formula: $2(Na + K) + (Glu)/18 + NPN/2.8$. Therapy is directed towards correcting the marked fluid deficit with HYPOTONIC fluids. Since these patients are more sensitive to the effects of insulin than are those in ketoacidosis, particular care must be exercised in drug therapy.

The presence of trismus on physical examination, particularly in an alcoholic derelict, should alert the physician to the possibility of alcohol-hypoglycemia (Ann Int Med 66:893 '67).

When the comatose person is noted to have minimal, bilateral papilledema, arterial blood gas determination should be included in the work-up. CO_2 retention increases cerebral blood flow and secondarily raises intracranial pressure causing papilledema.

In the comatose individual, pressure over the supraorbital nerve may cause sufficient pupillary dilatation for an adequate funduscopic examination. Pressure in this region is also a useful pain stimulus to test the depth of coma.

Comatose individuals with herniation of the hippocampal gyrus through the tentorial notch may be found to have the clivus ridge syndrome, the features of which are:
1. unilateral fixed mydriasis
2. oculomotor palsy
3. slight ptosis

Patients with long standing severe congestive heart failure at times have hypoglycemia (secondary to severe chronic passive congestion of the liver). In addition to coma, presentation of this condition may be bizarre or even psychotic behavior, palpitations, or sweating (in other words, findings which are often ascribed to anoxia, cerebro-vascular insufficiency, or electrolyte imbalance (NEJM 247:745 '52).

Miosis with coma may be due to pontine hemorrhage as well as to narcotics or sedative overdose.

CONJUNCTIVA

Yellow brown pigmentation of the exposed bulbar conjunctiva, which may appear as a well demarcated horizontal, 'icteric' band, is a characteristic photosensitivity reaction to large doses of phenothiazine drugs.

Localized hyperemia of the exposed conjunctiva is a common finding in alcoholic derelicts and may indicate multiple B vitamin deficiencies. Another sign of vitamin deficiency is "circumscribed thickening of the nasal bulbar conjunctiva," which is appreciated as a set of parallel vertical folds, concentric with the limbus, appearing on that surface during ocular adduction (Lancet 1:52 '68). Xerosis (dryness) of the conjunctiva is classically associated with vitamin A deficiency and in infants may be a harbinger of keratomalacia.

Brilliant red conjunctivitis or episcleritis is seen with and may be a presenting feature of Weil's disease and:
1. systemic lupus
2. polyarteritis
3. drug reaction
4. brucellosis
5. relapsing polychondritis
6. polycythemia

Conjunctival dermoids are not rare. They are usually located at the limbus and are often best recognized by the protrusion of lanugo hairs.

With chronic unilateral conjunctivitis in an elderly person, Meibomian gland carcinoma should be considered in the differential.

Conjunctival and episcleral blood vessels are best evaluated at the bedside with the hand ophthalmoscope set to the +40 lens, with the green light source. Some changes of note are:

1. The Paton sign - minute, comma-shaped vessels (present in almost 100% of persons with sickle cell anemia and 80% of those with S-C hemoglobinopathy, but much less than 20% of those with the sickle trait) (Arch Ophth 68:627 '62), typically, most densely located in the interior bulbar conjunctiva. *

* In children younger than 11 years, topical application of a mild vaso-constrictor may be necessary to make commas visible (AJO 69:563 '70).

2. Angulation, aneurysmaldilitation, and "sausage link" venous irregularities, characteristic of "Fabry's disease"; venous caliber irregularities are found with hyperviscosity syndromes (i.e. dysproteinemias, a specific finding is present for cryoglobulinemia, see "GLOBULINS").

3. True aneurysms of the conjunctival vessels in adults are often seen near the limbus and are common with significant arteriosclerosis and with chronic liver disease; conjunctival microaneurysms are not helpful towards a diagnosis of diabetes.

Calcium phosphate microcrystals are deposited in the external bulbar conjunctiva in more than 2/3 of cases of acute or chronic renal failure, giving a characteristic (and diagnostic) granular appearance. Crystalline forms are seen with the slit lamp. An inflammatory reaction to the crystals causes the "red eyes of renal failure" (Lancet 1:4 '67, 2:366 '68).

Pingueculae in a child suggests Gaucher's disease. Biopsy of a pingueculum at any age provides the definitive diagnosis (and with much greater safety than needle biopsy of an abdominal organ). A yellow pigmentation in a 'bat-wing' distribution over the face also occurs in some cases of Gaucher's disease.

CORNEA

In evaluating corneal opacity, consider:
1. acute or congenital glaucoma
2. keratitis (of any cause, i.e. herpes simplex, herpes zoster, bacterial infection, etc), chemical injury (esp. lye burns), radiation
3. corneal dystrophies
4. Hurler's syndrome (and other mucopolysaccharidoses)
5. old inactive congenital syphilis
6. Hand-Schüller-Christian disease

Corneal vascularization reportedly occurs with riboflavin deficiency. Colliquitive necrosis (sudden whitening, melting, and perforation) occurs with vitamin A deficiency.

Punctate staining of the cornea with fluorescein can be seen with systemic lupus (NEJM 276:1186 '67) as well as with:
1. Sjogren's syndrome
2. (other) causes of keratitis sicca (including the post-menopausal state, pemphigus, Stevens-Johnson syndrome, and damage to the lacrimal apparatus)
3. local injury or infection
This finding may account for the photophobia of some patients with SLE.

Arcus senilis is of no diagnostic significance in Negroes, although in Caucasians below age 40 it may indicate familial hypercholesterolemia. Arcus senilis begins superiorly and inferiorly in the corneal stroma in contrast to Kaiser-Fleisher rings which begin in the interpalpebral region (and which represent copper deposition in Descemet's membrane).

Angular, sparkling crystals in superficial corneal layers (and at the limbus) are virtually pathognomonic of cystinosis. In this disorder, crystals may also accumulate in the conjunctiva which may be biopsied for definitive diagnosis. Corneal crystal deposition is rarely, also, seen with multiple myeloma.

CYTOID BODIES

Cytoid bodies (="cotton-wool" exudates) represent micro-infarctions in the nerve-fiber layer of the retina and can be found in any condition in which there is localized retinal arteriolar ischemia or obstruction:
1. hypertension, toxemia of pregnancy, pheochromocytoma
2. severe anemia, circulatory collapse
3. collagen vascular disease (especially lupus)
4. diabetes
5. carotid artery occlusion or insufficiency
6. bacterial endocarditis, septicemia
7. serum sickness, anaphylaxis
8. fat embolization

9. dysproteinemias, hyperviscosity syndrome
10. carcinomatosis, lymphoma, leukemia (minute tumor emboli)

The incidental discovery of numerous cytoid bodies without other gross retinal pathology (in a woman) is most suspicious of subclinical lupus erythematosus.

DEAFNESS

Congenital deafness +
1. progressive nephritis = Alport's syndrome
2. goiter = Pendred's syndrome
3. retinitis pigmentosa = Usher's syndrome
4. syncope and (EKG) QT prolongation = Jervell-Lange syndrome

Congenital deafness occurs in some 20% of cases of Waardenburg's syndrome, an autosomal dominant with variable penetrance and expressivity characterized by:
1. teleocanthus (i.e. intercanthal distance greater than 55% of the interpupillary distance), ectopia of the lacrimal puncta
2. high broad nasal root, synophrys (eyebrows growing together)
3. white forelock, white lashes
4. heterochromia, depigmented irides (heterochromia may be extreme enough that there is one brown and one blue iris)
5. (variable) cleft lip or palate

DEPRESSION

Personality changes with refractory depression can be a presenting feature of as much as 5% of cases of pancreatic carcinoma. It is also a prominent feature of:
1. acute intermittent porphyria
2. hyperparathyroidism
3. Cushing's disease
4. collagen vascular disease
5. steroid therapy (see "STEROIDS")

Plutarch:

(The depressed man) looks upon himself as a man whom the gods hate and pursue with their anger...The physician, the consoling friends are driven away. 'Leave me' he says, 'me, the impious, the accursed, hated of the gods, to suffer my punishment'.....Awake he makes no use of his reason, and asleep, he enjoys no respite from his alarms.

DIAPHRAGM

Irritation of the diaphragm is associated with reflex soreness and tenderness to pressure of the upper border of the trapezius muscle on the same side. This finding is commonly present with diaphragmatic myocardial infarction and with pericarditis when the contiguous region of pericardium is inflamed. It is also found with subdiaphragmatic, splenic, or hepatic abscess and infiltrative pancreatic carcinoma.

Cutaneous hyperesthesia (or referred pain) of the shoulder or trapezeal region after 10 to 15 minutes in the Trendelenburg position (Kehr's sign) is a reliable indication of free blood in the abdominal cavity.

Total diaphragmatic excursion (obtained from inspiratory and expiratory chest films) appears to be proportional to lung compliance. Normal excursions range from 2.5 to 4.5 cm. This distance is increased with sarcoidosis and decreased with chronic obstructive pulmonary disease in which case severity of airway obstruction is indicated. (See Thorax 24:218 '69).

One physical sign of diaphragmatic pleurisy is pain on pressure at Gueneau de Mussy's point (which is located by extending lines from the left sternal border vertically and the end of the 10th rib horizontally.)

An elevated immobile diaphragm which cannot be explained raises the suspicion of pulmonary embolus.

DIARRHEA

Diarrhea and episodic hypertension in children is a presentation of NEUROBLASTOMA.

Diarrhea following milk intake is the presentation of lactose intolerance (lactase deficiency). Adult presentation is the rule. In a JHH study of male prisoners (Bayless): 19/20 Negro males had milk intolerance, but only 2/20 white males were lactase deficient.

Laxative abuse should be considered in the differential diagnosis of diarrhea in middle life, particularly in food fadists and in women preoccupied with bowel function. Frequently patients will not admit to excessive laxative use (or this material will not be sought in the history).

1. Prolonged fluid and potassium losses in the stool lead to dehydration and hypokalemia. As a result, this disorder may mimic Addison's disease, diabetes insipidus, K-losing nephritis, or colitis at the onset.

2. The diagnosis is suggested by the barium enema pattern of dilatation and atony of the ascending and transverse colon with loss of haustral markings and a patulous ileocecal valve (Lancet 1:1121 '66). Although loss of mucosal markings may suggest ulcerative colitis in a single view, the involved regions are not rigid and are seen to vary in shape in other views, differentiating these two causes of diarrhea.

Ice cold fluids should be restricted in patients with Crohn's disease, for they often precipitate diarrhea

DYSPHAGIA

Dysphagia is rarely seen with functional GI tract disease and therefore demands a complete work-up.

Esophageal carcinoma should be suspected when weight loss is out of proportion to dysphagia. Cytological examination of gastric washings may provide the diagnostic clue.

Diffuse esophageal spasm occurs with old age and with esophagitis. The clinical picture in this instance is dysphagia with recurrent lower sternal region pain accentuated by cold fluids and gastric reflux. Antacids and anticholinergics are the usual form of therapy.

Dysphagia can be seen with gastric ulcer as well as an invasive gastric malignancy. In the former case, it should revert within a few days of commencing intensive medical therapy.

When dysphagia is secondary to pharyngeal muscle dysfunction or weakness, there is usually regurgitation into the nose.

Cricopharyngeal muscle hypertrophy is a cause of painless inability to swallow solids referred to the region of C-6.

With achalasia or hiatus hernia, dysphagia may be unrecognized by the patient, who may present with pneumonia or recurrent pneumonitis secondary to aspiration during sleep.

A possible cause of dysphagia in elderly people (usually women) is compression of the lower, posterior esophagus by a dilated, atherosclerotic aorta (see Thorax 24:32 '69). Amyloid deposition in the lower esophagus may cause dysphagia.

(EXTERNAL) EAR

Increased pigmentation occurs with:
1. ochronosis
2. Addison's disease
3. hemochromatosis
4. thyrotoxicosis
5. scleroderma
6. Whipple's disease
7. chronic arsenic poisoning
8. MSH secreting tumors

The bluish ears of ochronosis are complemented by gray-black sclerae; scleral pigmentation is often interpalpebral, but characteristically (and diagnostically) it does not extend to the limbus (remaining anterior to the insertions of the horizontal eye muscles). Black droplets in the corneal stroma at the limbus are diagnostic!

Protruding ears are associated with pernicious anemia and with Werner's syndrome.

Inflexible ear cartilages are, rarely, seen with Addison's disease and after repeated direct trauma (i.e. prize fighters) or frostbite.

Libman spots (punctate red opacifications detected by transillumination of the ear lobe) indicate subacute bacterial endocarditis.

Ear lobe capillaries 'filter' large circulating cells. Increased histiocytes are noted in ear lobe blood from patients with SBE. Metastatic malignant cells can at times be found in buffy coats prepared from ear lobe blood.

Red ears and red eyes (i.e. episcleritis) indicate relapsing polychondritis. Involvement of the nasal septum leads to a saddle nose deformity. Other features are irregularities of the costochondral junctions and laryngeal involvement (which, when acute, may require tracheostomy). Intra-articular joint cartilage is not involved. There appears to be an association between relapsing polychondritis and dissecting aneurysm of the aorta.

Chronic otitis media, which does not respond to antibiotic therapy, can be a presenting complaint of Hand-Schuller-Christian disease, which is better known by the classical triad of:
1. exophthalmos

2. diabetes insipidus
3. defects in membranous bones

Very often, all of the features of the triad are not present. Proptosis is frequently unilateral early in the course. Erosion of the tooth containing region of the mandible is common and leads to loosening and displacement of the teeth and gingivitis, another presenting complaint which at first appears 'trivial'. In addition to frank otitis, a polyp in the auditory meatus of a child should raise suspicion of this disorder.

(EXTERNAL) EYE (including eyelid and lacrimal apparatus)
(See also "CONJUNCTIVA," "IRIS," "PROPTOSIS," "PUPIL,"
"NYSTAGMUS")

Slow or incomplete lid opening with rapid blinking is a sensitive indication of a neural conduction defect involving the lid musculature. This finding in an otherwise 'normal' individual predicts diabetes (Walsh, JHH).

Twitching of the eyelids with upward or extreme lateral gaze is Cogan's twitch sign of myasthenia (Arch Ophth 74:217 '65). It is abolished with Tensilon.

Plexiform swelling of the eyelid indicates neurofibromatosis (and is also Romaña's sign of Chagas' disease).

Tender painful swelling of the lacrimal gland occurs with mononucleosis and mumps. Painless lacrimal swelling is seen with sarcoidosis, Sjogren's syndrome, primary tumor, and lymphoma-leukemia (of this last category, this finding is most characteristic of lymphosarcoma).

Unilateral eyelid dermatitis on the side of the dominant hand in a woman suggests a contact reaction to nail polish. Patchy lid dermatitis is common with contact reactions to cosmetics, perfumes, scalp preparations, and hair sprays.

Unilateral ptosis with a dilated pupil may be due to a temporal lobe lesion, while ptosis with a miotic pupil if not due to Horner's or Raeder's syndromes, indicates a midbrain lesion in the region of the posterior commissure.

Delayed eye movements with inability to follow the target promptly when testing the extraocular muscles occurs with parietal lobe damage (Balint's sign). Impaired voluntary eye movements (with less than full excursions) are a feature of ataxia telangiectasia.

White eyelashes or eyebrows in a young person suggests the Vogt-Koyanagi syndrome, the other features of which are:
1. poliosis
2. bilateral uveitis
3. vitiligo (face, neck, or upper trunk)
4. alopecia areata
5. tinnitus or decreased auditory acuity

This syndrome was initially described in Japanese individuals but does occur rarely in occidentals; the association of bilateral uveitis and asceptic meningitis, which was formerly believed to be a variant of this syndrome, is now classified as Harada's syndrome.

Bilateral internuclear ophthalmoplegia is practically pathognomonic of multiple sclerosis (unilateral internuclear palsies occur most often with basilar vertebral vascular ischemic changes).

Heliotrope coloration of the eyelids diagnoses dermatomyositis.

Tightness of the eyelids can suggest scleroderma as a diagnosis, especially if there is difficulty everting the upper lid at examination (Paton, JHH).

Edema of the eyelids is the earliest sign of the muscular invasion stage of trichinosis, which occurs about a week after the infestation. Fever is present throughout this stage, eosinophilia begins about the 10th day. Later eye findings are:
1. conjunctival chemosis (particularly over the medial and lateral recti), subconjunctival hemorrhage
2. proptosis
3. muscle palsies
4. optic neuritis
5. retinal edema and hemorrhages

Upper lid retraction is usually due to hyperthyroidism (Dalrymple's sign), though it sometimes occurs in a setting of hepatic cirrhosis (NEJM 266:1244 '62) and is rarely a sign (Collier's) of a midbrain lesion.

With fracture of the anterior cranial fossa there may be circular lid ecchymosis, which is characteristically limited sharply by the orbital margin. Subconjunctival hemorrhage (which, therefore, does not move synchronously with ocular motion) may also occur. The 'black' eye of direct trauma, on the other hand, is not limited sharply by the orbital rim and may be associated with intraconjunctival hemorrhage. When ecchymosis involves both external eyes, and there is a history of a single blow to the head, basilar skull fracture is to be ruled out.

FACE

A dirty (or unshaven) face in an otherwise well-groomed individual suggests trigeminal neuralgia. Apropos, trigeminal neuralgia (or facial formication) in a young person should at least raise the suspicion of multiple sclerosis. Trigeminal neuralgia is not an uncommon concomitant of malaria.

Flushing:
pink - mastocytosis
purple - carcinoid (see "Carcinoid Syndrome")
plum - polycythemia vera
cherry red - carbon monoxide intoxication

Afternoon flushing is seen in tuberculosis. Unilateral flushing indicates lobar pneumonia but is also seen with apical tuberculosis and cervical sympathetic derangements.

Ruddy, malar flush, plethoric lips, and sallow perioral and frontal regions comprise the mitral facies of severe mitral stenosis and constrictive pericarditis.

prune face = pseudohypoparathyroidism
'moroccan' leather face = pseudoxanthoma elasticum
leonine face = leprosy, acromegaly

Periorbital (and perifollicular) purpura is characteristic of amyloidosis. Periorbital edema and muscle tenderness occur with dermatomyositis as well as trichinosis.

Unilateral lower lip weakness (becoming apparent in infants during crying) is an uncommon but apparently valid indication of some forms of congenital cardiac disease, particularly ventricular septal defect. This is the cardio-facial syndrome (J Ped 73:953 '68).

Facial myokymia (continuous, undulating, fine fascicular twitching of the facial muscles) may be an early indication of multiple sclerosis (Brain 84:31 '61). Facial myokymia usually begins with the perioral muscles. Myokymia restricted to the eyelids is common and indicates psychological stress rather than any systemic condition.

The appearance of acne in an elderly person raises the possibility of bromide use. Acneform eruptions may also follow B_{12} injections, the use of iodides (in expectorants), and the use of antituberculous drugs. Some common non-prescription bromide-containing preparations are Nytol, Bromo-Seltzer, and Nervine.

The commonest cause of parotid enlargement in adults is alcoholism. Other possibilities are:
1. sarcoidosis
2. infiltrative neoplasia, primary tumor
3. iodide hypersensitivity
4. starch ingestion (a habit of many Negro women)

Shakespeare (Henry VI, part 2, III:2):

Warwick (about the death of Gloucester)-

See how the blood is settled in his face.
Oft have I seen a timely parted ghost,
Of a shy semblance, meager, pale, and bloodless,
Being all descended to the laboring heart
Who in the conflict that it holds with death,
Attracts the same for aidance 'gainst the enemy,
Which with the heart there cools, and ne'er returneth
To blush and beautify the cheek again.
But see, his face is black and full of blood.
His eyeballs further out than when he lived,
Staring full ghastly like a strangled man;

FINGER NAILS (See also "CLUBBING," "SPLINTER HEMORRHAGES")

Shape changes:
1. flat - hepatic cirrhosis (Lancet 2:248 '51)
2. spooned (koilonychia) - iron deficiency anemia, congenital, and collagen-vascular disease, polycythemia vera, acromegaly, thyroid disease, Banti's syndrome (Texas State J M 61:620 '65)

Color changes:
1. yellow (Samman and White nail) - lymphatic hypoplasia (see "LEG")
2. black - ochronosis, mercury intoxication
3. green - B. pyocyaneus infection of the nail plate
4. blue-gray - methemoglobinemia

The nail bed may also be discolored as a drug reaction, i.e.:
1. blue-black (associated with pretibial and palatal pigmentation) - chloroquine (Arch Derm 88:419 '63)
2. purple-black - phenothiazine photo-reaction
3. brown (with peripheral oncholysis) - tetracycline photo-reaction (Arch Int Med 112:165 '63)

Whitening of the nail bed from the lumula to the periphery,

forming a white nail with a small irregular distal pink zone (the Terry nail) is a frequent finding with hepatic cirrhosis (Lancet 1:757 '54) which can also be found with rheumatoid arthritis and with diabetes.

Nails which are proximally white and distally red, with a sharp central demarcation (Lindsay's half and half nails) are most indicative of chronic renal disease with significant azotemia (Arch Int Med 119: 583 '67).

Half-moon (lunular) changes:
1. red - right sided congestive heart failure
2. blue - Wilson's disease (JAMA 166:904 '58)
3. brilliant blue - argyria (argyrol nose drops are still used by some elderly persons; a slate gray facial color is characteristic and should not be mistaken for cyanosis).

Lunulae are very often absent from all the fingers of patients who are cachectic or who have dermatomyositis.

Absence of lunulae of the thumbs with preservation on the other fingers (Banyai-Caden sign) occurs at times with active pulmonary tuberculosis (Arch Derm 48:306 '43).

Transverse white bands (Mees' lines) are typical of arsenic or thallium exposure (some 4 to 6 weeks earlier.) They may also follow acute tubular necrosis (Arch Int Med 117:276 '66).

Multiple transverse white bands (Muerhcke lines) are a reliable sign of hypoalbuminemia and are frequently found with hepatic disease and the nephrotic syndrome (Brit Med J 1:1327 '56).

Longitudinal pigmented bands are of no diagnostic significance in Negroes, although in whites that may be found with Addison's disease and with the Peutz-Jeghers syndrome.

Longitudinal ridging and beading of the nail substance appears to be a concomitant of aging, although it is frequent in younger individuals with rheumatoid arthritis (Ann Rheum Dis 19:167 '60).

Nail pitting occurs with acute rheumatic fever and is extremely common with active pulmonary tuberculosis. 'Geographic' pitting is present in about a third of individuals with psoriasis.

Splinter hemorrhages are present in 10 to 20% of the general population. They are particularly common in manual laborers (and are almost invariably found in professional dish and glassware washers). They are a classical sign of bacteremia (particularly that of bacterial endocarditis) and also occur with:
1. drug reactions
2. collagen vascular diseases
3. trichinosis
4. psoriasis
5. rheumatic valvular heart disease (without SBE)
6. thyrotoxicosis

GALACTORRHEA

Non-puerperal galactorrhea occurs with sedative-tranquilizer drug use (particularly meprobamate). It also occurs with:
1. pseudocyesis
2. herpes zoster
3. hypothalamic and ovarian tumors

4. Chiari-Frommel and Forbes-Albright syndromes
5. acromegaly

GENERAL APPEARANCE (See also "FACE")

Hippocrates:

> (The physician) should observe thus in acute diseases: first the countenance of the patient, if it be like those of persons in health, and more so, if like itself, for this is the best of all; whereas the opposite to it is the worst, such as the following: a sharp nose, hollow eyes, collapsed temples; the ears cold, contracted, and their lobes turned out; the skin about the forehead being rough, distended, and parched; the color of the whole face being green, black, livid, or lead colored.

Shakespeare (Henry V, II:3):

> Hostess (about Falstaff on his deathbed)-
>
> .... He is so shak'd of a burning quotidian tertian that it is most lamentable to behold.... I saw him fumble with the sheets, and play with flowers, and smile upon his finger's end, I knew there was but one way; for his nose was as sharp as a pen, and a babbl'd of green fields... So a' bade me lay more clothes on his feet. I put my hand into the bed and felt them, and they were cold as any stone.

GYNECOMASTIA

Gynecomastia occurs during estrogen therapy, with hepatic disease, and during the 'refeeding' phase of recovery from malnutrition. It is also seen with:
1. chronic renal disease
2. dialysis therapy for chronic renal disease
3. hyperthyroidism (usually toxic diffuse goiter)
4. digitalis, INH, or phenothiazine therapy

HANDS (See also "FINGER NAILS")

Palmar erythema occurs with hepatic disease and:
1. pregnancy
2. rheumatoid arthritis
3. beri-beri
4. arsenic intoxication
5. dermatomyositis

Minute finger tip scars (secondary to skin infarctions) are common in patients with systemic lupus or other causes of Reynaud's Phenomenon. With chronic finger tip ulcerations, allergic contact reaction should be considered (particularly to chrome). A social history may be enlightening if the patient works with leather, blue-prints, or cement regularly.

"Ice-pick" hands indicate BASAL CELL NEVUS SYNDROME.

Roughening and advancement of the cuticle over the nail substance (pterygium formation) suggests dermatomyositis or scleroderma. In the setting of a multi-system disease, this finding militates against the diagnosis of systemic lupus.

Interosseous muscle wasting is frequent in diabetes.

Hand "signs" of dermatomyositis are:
1. violaceous discoloration over the knuckles
2. telangiectases in the finger skin folds and along the nail margins
3. absence of the lunulae from all fingers

Brachydactyly may be associated with congenital cardiac obstructive outflow disorders, usually of the right side of the heart (Cardiologia 50:330 '68). Vestigal or finger-like thumbs are seen in association with cardiac septal defects (NEJM 272:437 '65). Syndromes in which there are hand abnormalities and congenital cardiac defects are (see NEJM ibid):
1. Holt-Oram syndrome
2. Down's syndrome
3. Trisomy D, Trisomy E
4. Gonadal dysgenesis (Turner's syndrome)
5. Ellis-van Crevald syndrome
6. Marfan's syndrome
7. Fanconi syndrome
8. Laurence-Moon-Biedel syndrome

Shortness of the fourth metacarpal (leading to prominence of the fourth knuckle) is a feature of Turner's syndrome. In addition to short stature, gonadal failure, and shield chest, other clinical features are:
1. webbing of the neck
2. triple (occipital) hair line
3. wide carrying angle of the elbow (cubitus valgus)
4. widely distributed pigmented nevi

Shortness of the lateral metacarpals (4th and 5th) is seen with pseudohypoparathyroidism, basal cell nevus syndrome, and some cases of hereditary multiple cartilagenous exostoses; shortness of the 5th metacarpal, congenital lues. Shortness of the middle phalanx of the 5th finger is common in mongolism.

HEADACHE

Retro-auricular headache with cutaneous hyperalgesia in the same region suggests acoustic neuroma. External auditory canal hyperesthesia is another 'early' sign of acoustic neuroma.

Persistent occipital headaches (radiating down the neck) raise the possibility of a subtentorial tumor. Lancenating bitemporal headaches are characteristic of a pituitary tumor.

Headache and confusion in an elderly person raises the possibility of a chronic subdural hematoma. When headache accompanies a cerebrovascular accident, hemorrhage is considerably more likely than thrombosis.

A rare form of headache, and one which indicates brain tumor, is paroxysmal, transient, incapacitating head pain. Brain tumor headache is usually intermittent and mild until late in the course. It may be aggravated by coughing, sneezing, stooping, straining at the stool, and upper respiratory infection (though this behavior is not specific). When there is no papilledema, unilateral headache is almost always ipselateral to the mass; indeed, when the headache is localized it often overlies the tumor.

Headache after alcohol ingestion is suggestive (though not specific) of Hodgkin's disease. Apropos, alcohol induced pain may occur at any site of lymphomatous infiltration.

Ophthalmoplegic migraine or migraine beginning in middle life raises the possibility of intracerebral aneurysm. Recurrent, severe, throbbing headache which is always on the same side of the head is characteristic of a circle of Willis aneurysm. Pre-retinal (subhyaloid) hemorrhage follows leakage of blood from a saccular aneurysm and when noted in a comatose patient is diagnostic of subarachnoid hemorrhage.

Combing the hair may be painful in a region involved with cranial arteritis and after a migraine headache.

HEART SOUNDS

When a very soft first heart sound is associated with a fourth heart sound, first degree A-V block is present.

In the upright position, splitting of the second sound in the pulmonic area is generally only audible during the inspiratory phase of respiration.

1. Expiratory splitting is always abnormal and implies either early aortic or late pulmonic valve closure
    right bundle branch block
    ASD with left to right shunt
    pulmonic stenosis

2. Fixed splitting indicates an inability to vary stroke volume of the right side, as with congestive failure. The sudden onset of wide, fixed splitting should suggest the possibility of massive pulmonary embolus.

3. Paradoxical splitting implies late aortic closure
    left bundle branch block
    severe aortic stenosis
or early pulmonic closure
    Wolff-Parkinson-White syndrome
    left ventricular aneurysm

Splitting of the second sound in the mitral area is an abnormal physical finding which occurs with severe pulmonary hypertension and with atrial septal defect when the right ventricle presents at the apex.

A single heart sound occurs when:

1. the right and left sides are subjected to similar pressure dynamics -
    the Eisenmenger complex

2. aortic and pulmonic closure sounds are superimposed -
    moderate aortic stenosis (see "AORTIC STENOSIS")

3. aortic closure is inaudible -
    severe calcific aortic stenosis

4. pulmonic closure is absent or inaudible -
    tetralogy of Fallot
    pulmonic atresia
    severe pulmonic stenosis

A tambouric second heart sound along the aortic outflow tract classically occurs with syphilitic aortitis; a "bonky" quality in this same region is indicative of atherosclerotic change.

An S4 occurs with mitral insufficiency of acute onset (such as papillary muscle dysfunction) but is generally not present with mitral insufficiency of long standing. When an atrial diastolic gallop is associated with a holosystolic murmur which has been present for some time, hypertrophic subaortic stenosis should be suspected (Spangler, U. Colo.).

HYPERVENTILATION

Unexplained respiratory alkalosis (i. e. hyperventilation) in patients with tumors or severe debilitating disease implies sepsis and requires prompt, intensive treatment.

In children, unexplained hyperventilation should raise the suspicion of salicylism. In infants, hyperventilation does not necessarily imply a pulmonary disorder in that relatively 'minor' degrees of vomiting or diarrhea may lead to metabolic acidosis and compensatory respiratory changes.

In tetany due to hysterical hyperventilation, the hand is characteristically clenched into a closed fist. Tetany does not occur with the hyperventilation induced as a compensation of metabolic acidosis.

Unexplained episodes of hyperventilation occur with decompensated hepatic cirrhosis. Dilutional hyponatremia is at times present (Am Rev Resp Dis 96:971 '67).

IMPOTENCE

Impotence may be a presenting feature of maturity onset diabetes or a late concomitant of hepatic-cirrhosis. It is also seen with:
1. postural hypotension, Leriche syndrome
2. temporal lobe lesions
3. multiple sclerosis, tabes dorsalis
4. therapeutic use of ganglionic blocking agent, bromism
5. extensive, though perhaps silent, intra-abdominal carcinoma
6. feminizing adrenal tumors, pituitary disorders, Addison's disease
7. thiamine, nicotinic acid, or B_{12} deficiency

Antihistamines (and the chemically related phenothiazines) at times interfere with ejaculation.

INSOMNIA

Insomnia in elderly patients may be secondary to decreased cerebral blood flow consequent to reduced cardiac output. Treatment of these cases should not be sedation.

Recurrent nocturnal anxiety or nightmares in elderly patients often precedes the occurrence of nocturnal dyspnea or nocturnal sweating, more obvious signs of cardiac decompensation.

When chronic brain syndrome is present, a light should be left on near the bed, despite insomnia. The disorientation which occurs in the darkness (i. e. when the major sensory stimulus is removed) leads to marked agitation.

Heterochromia may be due to:
1. congenital defect, poor ocular prosthesis
2. congenital Horner's syndrome (involved side lighter)
3. Fuch's heterochromic iridocyclitis (or other unilateral inflammatory process)
4. siderosis oculi (following an intraocular ferrous foreign body)
5. Waardenburg's syndrome
6. syringomyelia
7. juvenile xanthogranulomatosis
8. ring melanoma of the iris

When only one iris is a very dark brown color, melanosis oculi is a possibility. This disorder involves unilateral melanin deposition in the anterior structures of the eye and appears to predispose the involved eye to subsequent malignant melanoma. Excessive melanin deposition in one iris also occurs with a congenital subcutaneous nevus of the periorbital region (i.e., nevus of Ota).

Sporadic (i.e., non-familial) aniridia is associated with Wilm's tumor (Arch Ophth 75:796 '66).

Iris vascular anomalies (which can be seen with the slit lamp) are a feature of myotonic dystrophy (AJO 69:573 '70).

LEG (and foot)

Hypertensive ischemic ulcers are usually posterolateral, stasis ulcers are usually medial. Satellite regions of necrosis of leg ulcers in any region suggest an hypertensive ischemic etiology. Midline posterior ulcers which are sharply marginated occur with the arteritis of collagen-vascular disorders.

Edema of the heel indicates an arthritic process, since dependent edema does not involve this region.

Irregular gout should be suspected with recurrent thigh and calf myalgias (despite normal uric acid levels), particularly if there is symptomatic relief with rest and prompt recurrence with standing (Am J Med 54:267 '61).

Unilateral edema with a bluish tint to the toes should raise the suspicion of Kaposi's sarcoma, particularly in persons of Jewish or Mediterranean ancestry.

Painful unilateral ankle swelling which is not accentuated by motion is a feature of sarcoid. Painful bilateral ankle and dorsal foot edema occurs with pyridoxine deficiency.

Purple discoloration of the sides and plantar surfaces of the toes which blanches on pressure is an unusual complication of coumadin therapy (Ann Int Med 55:911 '61). These changes later extend to involve the greater portion of the foot. This complication does not imply that drug dose is excessive as regards its effect on clotting.

When leg edema is not obviously due to cardiac, renal, or nutritional causes, the finding of slow growing, brittle, yellow nails (Samman and White nails, Brit J Derm 76:153 '64) indicates lymphatic hypoplasia. Yellowing may precede edema by some time. Although the legs are generally involved, recurrent pleural effusions (secondary to defective pleural lymphatic drainage) may also occur.

Leg ulcers may be due to arterial insufficiency and:
1. Sickle cell disease, congenital spnerocytic anemia (especially over the malleoli in these two states)
2. lupus, scleroderma
3. rheumatoid arthritis
4. diabetes (pigmented shin spots)
5. ulcerative colitis
6. pellagra
7. self-infliction

Painless weakness of the legs precipitated by exertion (and associated with a heavy, numb feeling) is the clinical picture of cord ischemia. During ischemia, deep tendon reflexes are brisk, the plantar reflex is extensor, and there may be loss of sphincter control. The weakness is generally asymmetrical. As with transient cerebro-vascular insufficiency, there are no neurological signs prior to the attack. With repeated ischemic episodes, the clinical picture becomes that of flaccid paresis which has the features of combined upper and lower motor neuron disease.

Increased heel pad thickness is a roentgenographic indication of the presence or absence of acromegaly (Brit J Radiol 43:119 '70).

Pitting edema of the foot shortly after birth without cardiac or renal disease is a feature of Turner's syndrome and results from lymphatic hypoplasia. Leg X-rays later on may reveal large medial femoral condyles (A J Roent 89:1222 '63).

Foot or ankle pain of recent onset with localized tenderness and low grade fever (in any age group) may be a manifestation of an occult streptococcal infection (Proc Soc Med 63:409 '70). Localized erythematous patches in a patient with the above triad add certainty to this clinical possibility; in any case, workup should include a sed rate and ASLO titer before the case is pronounced a local orthopedic dysfunction. Penicillin is therapeutic.

LENS

Lens discoloration may be idiopathic or occur with:
1. Marfan's syndrome
2. homocystinuria
3. Marchesani's syndrome (spherophakia)
4. trauma
5. syphilis
6. coloboma, congenital dislocation

With homocystinuria the lens is usually displaced DOWNwards, with Marfan's syndrome it is usually displaced UPwards.

Cataracts with red, blue, and green crystalline deposits (i.e. polychromatic cataracts) occur with endocrinopathies and with myotonic dystrophy.

About 10% of patients with atopic eczema develop cataracts, characteristically stellate, anterior, subcapsular, shieldlike opacifications. The incidence of retinal detachment is also increased in this group of patients, although not as much as are cataracts.

Epitrochlear nodes are especially prominent with mononucleosis and with lues.

Occipital nodes are enlarged with roseola infantum; concurrent, prominent enlargement of retro-auricular, post-cervical, and pre-occipital nodes is characteristic of German measles.

The posterior cervical nodes are often enlarged with infectious hepatitis and early in the course of mononucleosis.

Pre-auricular nodes may be involved with parotid tumors, scalp neoplasms, and local granulomatous or infectious processes of the face, particularly the periocular and orbital regions. Painless pre-auricular node enlargement in times past, however, was said to be particularly suspicious of tuberculosis; careful examination of the undersurface of the upper eyelid on the involved side is always indicated in this situation.

Enlargement of the nodes behind the clavicular insertion of the sternocleidomastoid muscle indicates lymphoma or invasive carcinoma. In order to examine these nodes, the patient should perform a Valsalva maneuver, which will raise these nodes sufficiently that they may be palpated from the median aspect of the muscle.

Unilateral hilar node enlargement is seen with bronchogenic and metastatic lung cancer and:
1. lymphoma, leukemia
2. fungal infections
3. very early sarcoidosis
4. childhood tuberculosis

Indeed, node enlargement is the rule in primary TB of childhood. Conversely, single or multiple pulmonary infiltrates with enlarged nodes on the chest film of a child should suggest TB as well as more ominous diagnoses. Tracheobronchial abnormalities such as narrowing or indentation are present in about 30% of cases and add to the suspicion of TB when other changes are present. Paradoxically, lymph nodes on chest film often become larger early in the treatment of childhood TB, at times causing segmental collapse.

Carinal (mediastinal) node enlargement is simply and reliably evaluated by chest film with barium swallow, the LAO position is optimal (AJR 90:792 '63). At times, hilar adenopathy is more clearly revealed by fluoroscopy or chest films taken some 5 to 10 seconds after the initiation of the Valsalva maneuver, when cardiac and pulmonary vascular size decreases.

The left supraclavicular - scalene node group should be biopsied in all women with cancer of the cervix, regardless of the clinical stage of the lesion. Although there may be no other indications of distant spread of this cancer, these nodes are often microscopically involved, a finding of therapeutic moment (Ketcham, NCI-NIH).

THE MOUTH AND JAW

Swollen gums:
1. pregnancy
2. dilantin intoxication
3. leukemia (particularly monocytic)
4. scurvy, beri-beri

5. macroglobulinemia
6. lung abscess

The "lead line" is a blue-black region of lead sulfide deposition along the gingival margins of the gums. A red line rimming the teeth may often accompany pulmonary tuberculosis (Fredericq's sign), a purple line indicates gold poisoning.

An enlarged tongue occurs with:
1. amyloidosis
2. lymphoma
3. myxedema
4. riboflavin deficiency
5. angioneurotic edema
6. superior vena cava syndrome
7. mercurialism

An enlarged, painful tongue which may become gangrenous, may occur with giant cell arteritis.

Hypersalivation may be psychogenic or related to:
1. distal esophageal carcinoma, peptic esophagitis
2. epilepsy
3. rabies
4. mercurialism
5. digitalis intoxication
6. anticholinergic agent intoxication (i.e. roach poison)
7. bronchial carcinoid tumor

Petechial hemorrhages can often be seen on the undersurface of the tongue when they are not found elsewhere.

Mucosal ulcerations are seen with local aphthous disease and:
1. diabetes
2. secondary syphilis
3. Reiter's syndrome
4. Behcet's syndrome
5. ulcerative colitis
6. agranulocytosis
7. Hand-Schuller-Christian disease
8. Stevens-Johnson syndrome
9. systemic lupus
10. histoplasmosis

Gangrenous oral cavity bacterial infections are a feature of the rare Takahara's disease (autosomal recessive acatalasia), which is of world-wide distribution (Lancet 2:1101 '52).

Intermittent claudication of the jaw is pathognomonic of giant cell arteritis (Tumulty, JHH).

The triad of: 1. furrowed (scrotal) tongue
2. recurrent VIIth nerve palsy (with lagophthalmos)and
3. swelling of the face (especially the lips)
is the Melkersson-Rosenthal syndrome.

NOSE

An unilateral, persistent, purulent discharge is most often caused by a foreign body. It may also be seen with the now uncommon occurrence of nasal diphtheria.

A perforated nasal septum is usually secondary to trauma but may also be caused by:

1. syphilis
2. collagen vascular disease with small vessel arteritis (i. e. SLE or polyarteritis)
3. aortic arch syndromes (pulseless disease)
4. Wegener's granulomatosis, midline granuloma
5. chromate poisoning
6. carcinoma
7. relapsing polychondritis

Neonatal stuffy nose follows obstetric use of reserpine; the differential diagnosis must include congenital syphilis and Hurler's syndrome.

When herpes zoster involves the nasociliary nerve (skin changes are noted at the tip of the nose), ocular involvement is certain.

Unilateral cranial nerve syndromes due to invasive nasopharyngeal carcinoma are:

1. Trotter's syndrome:
   deafness
   mandibular V neuralgia (with lower jaw pain)
   palatal paralysis

2. Godtfredsen's syndrome:
   trigeminal neuralgia
   ophthalmoplegia (VI palsy)
   hypoglossal palsy

3. Garcin's syndrome:
   (unilateral, global cranial nerve paralysis without increased intracranial pressure of cerebral signs

4. Jacod's syndrome (petrosphenoid):
   total ophthalmoplegia
   amaurosis
   trigeminal neuralgia

5. Pterygopalatine fossa syndrome:
   amaurosis
   deafness
   infraorbital anesthesia, upper jaw pain
   pterygoid muscle paralysis

Sneezing is common during the premonitory phase of measles and characteristically increases after the initial fever (before the rash appears). Indeed measles should be kept in mind when a youngster presents with a running nose or other cold symptom etiology particularly when there is also some degree of photophobia. Koplik spots on the buccal membrane are, of course, diagnostic (and should be looked for in sunlight if possible). They may also be noted on the conjunctival caruncle or along the semilunar folds. Warthin-Finkeldey cells are present in pap smears of the nasal mucus during the prodromal phase, a diagnostic finding (JAMA 157:711 '55).

NYSTAGMUS

Vertical nystagmus specifically localizes the lesion to the brain stem, although it must be recalled that in addition to extrinsic pressure effects and intrinsic stem lesions, vertical nystagmus can, and indeed most often is, caused by certain drugs, namely intoxication with barbiturates, diphenylhydantoin, or ethanol.

If in a middle ear infection, the direction of the nystagmus changes (i. e. the direction of the rapid component shifts to the opposite side), the infection has spread and a cerebellar abscess has formed.

Ocular myoclonus should not be mistaken for nystagmus; it is a rare but characteristic sign of vertebro-basilar insufficiency. Ocular 'bobbing' (Arch Ophth 82:774 '69) is a spontaneous, recurrent (though not necessarily rhythmic) rapid depression of the eye followed by a slow return and is an ominous sign of brain stem or pontine damage.

Nystagmus in all directions is a feature of barbiturate intoxication.

OLFACTION

Increased olfactory (and other special sense) acuity occurs with Addison's disease and with virilizing, congenital (nonhypertensive) adrenal cortical hyperplasia. Decreased special sense acuity is regularly noted with myxedema (though the mechanism in this case is probably central). Hyposmia is a clinical feature of pseudohypoparathyroidism (J Clin Endo 28:624 '68).

Olfactory acuity appears to depend to some extent on vitamin A activity. Reversible hyposmia is noted with experimental vitamin A deficiency, cases of malabsorption (with low A and carotene levels), and acute viral hepatitis at the height of the disease process (Henkin, NIH).

Hyposmia is a regular (perhaps even constant) complication of laryngectomy (Lancet 2:479 '68). It is also noted with various forms of hypogonadism and following head trauma (especially with fracture of the cribriform plate).

Volatile irritants such as ammonia affect the sensory endings of the trigeminal nerve in the nose as well as the olfactory sensors. Therefore, when there is anosmia to ammonia (and the nasal mucosa is intact), hysteria is likely.

PROPTOSIS (See also "EXTERNAL EYE")

The commonest cause of proptosis in orientals is nasopharyngeal carcinoma. The commonest primary orbital tumor in adult occidentals is a cavernous hemangioma; in children, dermoid cyst.

Proptosis in a patient with von Recklinghausen's disease usually indicates an optic glioma, though if the eyelid is involved, a plexiform neuroma is likely.

When proptosis is associated with fullness of the temporal fossa, sphenoid ridge meningioma is to be ruled out (Am J Ophth 45:30 '58).

When proptosis in a child occurs with spontaneous ecchymosis of the eyelid, think of metastatic neuroblastoma.

Proptosis and a conjunctival salmon spot suggests lymphoma.

Pulsatile exophthalmos is practically pathognomonic of carotid-cavernous sinus fistula (which may follow relatively minor head trauma in elderly individuals with carotid atherosclerotic disease). Other physical findings are a bruit over the temporal-frontal zone, conjunctival congestion (or chemosis) and a III or VI cranial nerve palsy. The retinal veins may be dilated. Essentially, the only other cause of pulsatile exophthalmos is neurofibromatosis with erosion of the posterior orbit and transmission of cerebral pulsations.

Caffey's disease (infantile cortical hyperostosis) should be suspected when proptosis appears during the first 6 months of life.

PRURITIS

Pruritis can be seen with:
1. diabetes
2. uremia
3. malignancy
4. atopic dermatitis, allergic reactions
5. "dry skin" - hypothyroidism, aging, hypervitaminosis A
6. hypercalcemia
7. hepatic disease - obstructive jaundice
8. psychosis, chronic brain syndrome

Pruritis intensified with hot baths is characteristic of polycythemia vera.

Pruritis occurs in about 20% of cases of Hodgkin's disease, and in this condition it is usually continuous and often associated with burning paresthesias; pruritis confined to the lower half of the body also suggests this diagnosis.

Transient pruritis after eating sweets is sometimes seen with latent diabetes.

(PERIPHERAL) PULSE

Harvey (1628):

A certain person was affected with a large pulsating tumor on the right side of the neck, called an aneurysm, just at that part where the artery descends into the axilla, produced by an erosion of the artery itself, and daily increasing in size; this tumor was visibly distended as it received the charge of blood brought to it by the artery with each stroke of the heart. The connection of parts was obvious when the body of the patient came to be opened after his death. The pulse in the corresponding arm was small, in consequence of the greater portion of blood being diverted into the tumor and so intercepted.
Whence it appears that wherever the motion of the blood through the arteries is impeded, whether it be by compression, infarction or interception, there do the remote divisions of the arteries beat less forcibly, seeing that the pulse is nothing more than the impulse or shock of the blood in these vessels.

A bounding pulse is seen with:
1. primary increase in cardiac output - aortic insufficiency, patent ductus arteriosis, single large A-V fistula, multiple small A-V communications

2. increased stroke volume - complete heart block, truncus arteriosus, tetralogy of Fallot

3. systemic conditions with secondary increase in cardiac output - thyrotoxicosis, anemia, pregnancy, dehydration, beri-beri, anxiety

With a bispherians pulse there are two peaks present during systole. This type of pulse indicates hypertrophic subaortic stenosis or combined aortic stenosis and insufficiency.

With a dicrotic pulse there are separate peaks in systole and diastole. It is best appreciated in the brachial arteries and represents a disparity between stroke volume and arterial capacity (Ewy, Georgetown). It is seen, for example, in primary myocardial disease or severe ischemic heart disease in which vascular resistance is high but stroke volume is limited.

A paradoxical pulse may be defined as a decline in systolic blood pressure of more than 20 mm Hg with regular inspiration. A palpable pulsus paradoxicus is always significant. It occurs with:
1. severe congestive failure
2. superior vena cava syndrome
3. chronic obstructive lung disease
4. pericardial effusion

When a paradoxical pulse is limited to the upper extremities, thoracic outlet syndrome must be considered (Swinton, Walter Reed).

A pulsus alternans indicates left ventricular failure, although it is not necessarily a very late sign of decompensation (Angiology 19: 103 '68).

Adams (Dublin Hosp Rep 4:353 1827):

....decided symptoms are... the peculiar irregularity or want of correspondence in the pulse, as felt at the wrist and examined simultaneously at the heart; the latter often beats so violently against the sides of the thorax as to shake the patient in his bed, while at the same time the arterial pulse is small, weak, and irregular.

The course or even presence of dorsalis pedis arteries is sufficiently variable that little significance can be attached to absent pulsations over the dorsum of the foot at the initial examination. Absence of posterior tibial pulsation, however, is always significant. The converse is not true, namely, that the presence of posterior tibial pulses rules out proximal arterial obstruction in that these pulses may only disappear when exercise precipitates claudication (NEJM 262:1214 '60).

PUPIL

Slight anisocoria and sluggishness of constriction to direct light is an 'incipient' Argyll-Robertson pupil. Failure to induce pupillary dilatation with painful skin stimulae is a constant feature of the AR pupil. The association of the AR pupil and paralysis of conjugate upward gaze (Parinaud's syndrome) indicates a midline lesion in the region of the colliculi, such as a pinealoma.

The Norris-Fawcett sign of increased intracranial pressure is pupillary dilatation after 10 seconds of flexion of the neck (Arch Neurol 12:381 '65). This sign is present in about half of patients with increased ICP due to mass lesions. The test, obviously, should not be performed when there are signs of brain stem dysfunction or suspicion of trauma to the cervical vertebrae.

Systolic pupil contraction and diastolic dilatation is Landolphi's sign of aortic insufficiency. Hippus (rhythmic variations in pupil size) is of little diagnostic significance, occurring in 'normal' individuals and those with:

1. developing cataracts
2. cerebral tumors
3. epilepsy
4. disseminated sclerosis
5. general paresis

Unilateral hippus often develops contralateral to a CVA with hemiparesis.

Day to day variation (or even reversal) of pupil inequality is seen with <u>central anisocoria</u> (pupil reactions are preserved). This finding occurs with multiple sclerosis (Walsh, JHH).

Miosis in association with lacrimation, salivation, rhinorrhea, and profuse sweating should suggest (in rural regions) acute mushroom poisoning and (in cities) anticholinesterase intoxication of insecticide ingestion (particularly roach poison). With both there is also gastric distress, nausea, vomiting, and abdominal pain, and for both heroic doses of atropine are mandatory. With roach poisoning there may also be nicotinic pharmacologic effects further complicating the presentation. Central or peripheral respiratory paralysis may occur and there is often wheezing secondary to bronchoconstriction; oxygen exchange is further compromised by increased bronchial secretions. A-V heart block may be present.

SKIN

Baer (Ann NY Acad Sci 123:354 '65):

Due to the amazing pluripotentiality of reactivity of the skin, the character of the reactions produced by drugs assumes many different forms and as a matter of fact can imitate many known skin diseases. With very few exceptions, then, the possibility of a drug reaction must be considered in diagnosing many of the eruptions which are seen clinically.

The occurrence of herpes zoster in an elderly patient requires exclusion of a spinal structure compression (or carcinomatous metastasis) with appropriate vertebral column X-rays. Apropos, vertebral collapse in an elderly woman raises the suspicion of breast cancer (or in either sex, lymphoma or multiple myeloma).

With an atypical psoriaform dermatitis, i.e. one with asymmetry or without prominent involvement of the usual sites of predilection (umbilicus, extensor forearm surfaces, retroauricular zones), consider cutaneous lymphoma.

When a papular eruption occurs on the palms (and/or non-weight bearing portions of the soles) but does not involve the scalp, elbows, or knees, it almost certainly is due to the secondary stage of the French disease.

'Phantom' spots of the scalp (regions of transient hyperesthesia, which may later ulcerate) occur with polyarteritis and with giant cell arteritis (Tumulty, JHH).

Some cutaneous manifestations of neoplasia are:
1. acanthosis nigricans (usually adenocarcinomas)
2. freckles and/or senile keratoses appearing suddenly in later

life (Leser-Trelat sign)
3. ichthyosis with hyperkeratosis of the palms and soles
(lymphoma)
4. epidermoid cysts of the face and extremities (part of
Gardner's syndrome)
5. perioral, buccal, or finger pigmentary lesions (part of the
Puetz-Jeghers syndrome)
6. dermatomyositis
7. pachydermoperiostosis
8. erythroderma
9. Herpes zoster

Colored striae in the young are an indication of puberty and reflect changes in steroid production. They are not necessarily related
to obesity or mechanical skin changes (J Ped 45:520 '54). Colored
striae in obese adults indicate increased corticoid production (NEJM
266:1031 '62).

Xanthomas secondary to obstructive jaundice rarely involve
tendon sheaths. When a patient with type II (beta) hyperlipoproteinemia
has tendon xanthomas, at least one of his parents will almost invariably
also have hyper-betalipoproteinemia (Fredrickson, NIH). The presence
of palmar lipid deposition in a patient with hyperlipoproteinemia indicates a type III (broad beta) abnormality.

When examining a patient for metastatic or recurrent malignant
melanoma, it is essential not merely to look at the skin but to feel all
of the skin (Pilch, NIH). Metastatic melanoma has the appearance of
blue tipped subcutaneous nodes; when the primary site is not evident,
the mucous membranes should be re-examined. Amelanotic melanoma
of the sole may be mistaken for a simple plantar wart (Chretian, NIH).

Vital staining of a skin lesion with (topical) toluidine blue indicates that the lesion is neoplastic (Arch Surg 100:240 '70). This test
is a valuable supplement to the initial clinical examination.

SKULL

Cushing (1928):

By a strange human frailty, auscultation of the skull seems to
be the one thing most likely to be neglected in a routine neurological
examination.

Cranial bruits are heard with vascular intracranial tumors
(such as an angioma), with carotid-cavernous sinus fistula, with
marked arteriosclerotic change of the internal carotid in the syphon,
and over bone severely involved with Paget's disease (through which
there is a functional A-V fistula). Although vascular flow bruits may
be present with any cause of increased cardiac output (see "CHEST"),
bruits over the eyes are strongly suggestive of hyperthyroidism.

Cranial bruit (over the anterior fontanelle or posterior temporal zone) has been reported in 82% of children (aged 3 months to 5
years) with purulent meningitis (NEJM 278:1420 '68).

Vocal resonance may be detected over osteolytic skull lesions
(NEJM 264:1203 '61). Apropos, it is well to recall that in addition to
multiple myeloma, metastatic carcinoma, and other neoplastic disorders, osteolytic lesions may be due to localized osteomyelitis (which
is amenable to therapy).

Bony bossing (or fullness) may overlie a meningioma (see "PROP-TOSIS").

Prominence and tortuosity of scalp vessels may accompany vascular intracranial tumors, usually on the same side as the tumor (and most often evident in the superficial temporal region). A helpful confirmatory sign of a vascular intracranial tumor (such as an angioma) is rapid and marked venous filling above a manually obstructed internal or external jugular vein in the lower neck (see BRAIN 83:425 '60).

SYNCOPE

Exercise induced syncope should suggest disorders which limit sudden increases of left ventricular output, namely:
1. aortic stenosis
2. pulmonic stenosis
3. pulmonary hypertension

(Exercise induced cardiac failure occurs with beri-beri.)

'Shaving' syncope in men indicates a 'sensitive' carotid sinus. In general, when a sensitive carotid sinus is the cause of syncope, symptoms can be reproduced by very LIGHT pressure over the sinus zone. If deep massage is required, another cause must be sought. (The mechanism of syncope involves both bradycardia and decreased peripheral vascular resistance).

TELANGIECTASIS

Telangiectases are well known features of hepatic disease, ataxia-telangiectasia, Osler-Weber-Rendu disease, and scleroderma. They are common during pregnancy and can also be seen with:
1. lupus erythematosus, dermatomyositis
2. Cushing's disease
3. Ehlers-Danlos syndrome
4. carcinoid syndrome
5. polycythemia vera
6. systemic mastocytosis
7. xeroderma pigmentosa
8. riboflavin deficiency
9. otherwise "normal" individuals (about 10% of the general population)

Spiders of liver diseases are located above the nipple line, those of ataxia-telangiectasia in 'atopic' regions (skin folds, ears, and, particularly, the canthi of the eyes). Those of dermatomyositis occur over joints (especially the knuckles of the hands) and on the eyelids.

The production of white spots in the skin of persons with hepatic disease with local cooling indicates sites where telangiectases will form.

The number of spiders correlates very roughly with the extent of liver necrosis (Iber, JHH).

(BODY) TEMPERATURE

Fever brings out or accentuates already present neurological findings. It is essential to prevent fever in patients with CNS disorders. Apropos, both heat and exercise impair visual acuity in patients with multiple sclerosis (Arch Ophth 72:168 '64).

Controlled hypothermia will be ineffective unless the shivering mechanism is paralyzed with a drug such as chlorpromazine. Cryoglobulinemia should be excluded before the hypothermic blanket is used (Dove, USPHS).

Patients with sickle cell disease experience more discomfort during cold weather. Lowered ambient temperature increases the viscosity of hemoglobin S blood, predisposing to stasis and thrombosis.

Shivering ceases below body temperature of 90°. Treatment of patients with prolonged low temperature environmental exposure (such as derelicts sleeping in doorways and on the streets during the winter) should consist of:
1. slow elevation of the body temperature
2. intravenous fluids (relative hypovolemia may occur as the peripheral vascular bed dilates)
3. constant monitoring for cardiac arrhythmias
4. antibiotics for the pneumonia which is almost invariably present

Myxedema leads the differential diagnosis of 'endogenous' hypothermia. Hypothermia is also a clinical feature of hypoglycemia.

Petersdorf & Bennett (Ann Int Med 46:1039 '57):

There are several clues that call attention to the possibility of spurious fever:
(a) failure of the temperature curve to follow the normal diurnal gradient of body temperature - higher in the late afternoon and early evening;

(b) absence of tachycardia in the face of abrupt spikes in temperature;

(c) strikingly rapid defervesence unaccompanied by diaphoresis

(d) presence of fever of 106° F. or higher, a relatively rare phenomenon in adults.

When there is suspicion that a fever is factitious, it may be enlightening to determine temperature unobtusively by collecting a fresh urine specimen in a darkened container in which a thermometer has been secreted (see Arch Int Med 64:800 1939).

Fever and photophobia raise the possibility of meningitis.

TESTIS

Tender and atrophied testes occur with polyarteritis nodosa. Indeed, testicular biopsy in this situation will often confirm the diagnosis.

When gynecomastia is secondary to Leydig cell hypertrophy, urine chorionic gonadotrophin is high.

The left testicular vein empties into the left renal vein (the right into the inferior vena cava). Accordingly, left testicular enlargement with scrotal edema will be found with left renal vein thrombosis or obstruction by tumor.

36

Small firm testes with otherwise normal male genitalia suggest Klinefelter's syndrome. Abnormal testicular development is frequently associated with failure of descent; indeed, some 10% of Klinefelter's patients have undescended testes. (In the school aged male, undescended testis must be distinguished from retracted testis, a benign self-limited disorder.)

The testes of patients with cystic fibrosis are at times fibrotic. However, absence (or atresia) of the vas deferens is a characteristic and extremely frequent finding (Arch Path 88:569 '69). In about a third of all males with cystic fibrosis there is atrophy of some or all of the epididimis.

Testes remain prepubertal (unless gonadotrophins are administered) with hypogonadism secondary to pituitary disorders and:

1. as an autosomal recessive in association with mental retardation, retinitis pigmentosa, obesity, polydactyly (Laurence-Moon-Biedl syndrome)

2. as an X-linked trait associated with agenesis of the olfactory lobes with resultant anosmia (Kallman's syndrome) (partial loss of smell may be found in the mothers of these patients).

VISUAL FIELDS

Bitemporal hemianopsia is classical for pituitary tumor, but a bitemporal hemianopsia with pronounced papilledema suggests a cerebellar or third ventrical tumor. Since sellar tumors rarely exert pressure directly in the midline, early on, field defects are asymmetrical and are often small and central. When a pituitary tumor is associated with papilledema, it is likely malignant.

Sudden onset of homonymous hemianopsia with concomitant brain stem signs indicates basilar artery insufficiency.

Changing visual field defects are most characteristic of aneurysms. The tumor most commonly causing changing fields is the craniopharyngioma (Walsh, JHH).

Homonymous hemianopsia occurs in about a third of cases of carotid occlusive disease or middle cerebral artery insufficiency. (Apropos, carotid occlusive disease may rarely present with ipselateral glaucoma, JAMA 182:683 '62). Homonymous hemianopsia with central sparing is characteristic of ischemia in the posterior cerebral artery distribution.

A "pie in the sky" field defect (contralateral homonymous superior wedge-shaped defect) is almost always indicative of a temporal lobe disorder (Walsh, JHH).

Papilledema and papillitis can be clinically differentiated in that the former is rarely associated with a visual field loss, whereas the latter almost certainly is. An enlarged blind spot is a visual field feature of papilledema.

VITILIGO

Vitiligo is associated with some systemic disorders and may reflect an 'autoimmune' process. These disorders are (Brit J Derm 80: 135 '68):

1. pernicious anemia
2. Addison's disease
3. thyroid disease (all types)
4. diabetes mellitus

In this situation, depigmentation is characteristically symmetrical.

Autoantibodies have been demonstrated in some cases of vitiligo without any apparent systemic disease concomitant (Lancet 2:177 '69).

Vitiligo of the chest indicates scleroderma (Tumulty, JHH).

DISEASES

Charcot:

Disease is from old and nothing about it has changed. It is we who change as we learn to recognize what was formerly imperceptible.

ADDISON'S DISEASE

Addison (1855):

.... general languor and debility, remarkable feebleness of the heart's action, irritability of the stomach, and a peculiar change in the color of the skin.

All special sense modalities are sensitized:
1. olfactory - J Clin Invest 45:1631 '66
2. taste - J Clin Endo 27:214 '67
3. hearing (especially the 1000-2000 cps range) - J Clin Invest 46:429 '67.

Inflexible ear cartilages (Thorn's sign) occur with some cases of Addison's disease, although the mechanism of ear cartilage calcification is not as yet known.

Acquired perilimbal pigmentation on the eyes can be the best index of increased pigmentation due to Addison's disease in Negro patients (Paton, JHH).

Lymphocytosis is usual, and the proportion of large, immature lymphocytes is increased. This lymphocyte left shift may at times be sufficiently marked as to suggest the diagnosis of a lymphoblastic disorder. Lymphocyte immaturity is reversed by replacement therapy, but lymphocytosis tends to persist (Am J Med 42:855 '67).

Bilateral adrenal metastases are not uncommon with adenocarcarcinoma of the lung.

Addison's disease associations:
1. hypoparathyroidism and moniliasis
2. pernicious anemia
3. Hashimoto's disease
4. hypothyroidism (Schmidt's syndrome)
5. diabetes
6. vitiligo, alopecia totalis

A loud voice militates against the diagnosis of Addison's disease.

Alcoholics very often present with the features of various B vitamin deficiencies. In addition to pellagrous skin changes and a beefy, depapillated tongue, other indications are corneal neovascularization, conjunctival folding, calf tenderness, and painful ankle or dorsal foot edema.

Tremulousness is associated with hyperventilation. Neurological symptoms of impending DT's appear to be aided and abetted by respiratory alkalosis. Therefore, arterial blood gas and pH determinations should be a part of management. CO_2 rebreathing may at times be therapeutic.

Acute brain syndrome appearing in alcoholics residing in rural regions raises the suspicion of lead intoxication in that automobile radiators and other lead-lined containers are sometimes used by moon-shine distillers.

Cerebral fat embolism from an hepatic source may be involved in the acute brain syndrome of alcoholism (Lancet 2:188 '69).

Pulmonary tuberculosis occurs with increased frequency in alcoholics who have had partial gastrectomy.

The extra-ocular muscle palsies of Wernicke's encephalopathy reverse within hours of intravenous thiamine administration. Intravenous glucose must not be given to alcoholics without parenteral thiamine coverage (preferential hepatic thiamine uptake will cause brain 'starvation' in the thiamine deficient patient in such a situation).

Silky hair, indistinguishable from that of hyperthyroidism is very common in alcoholics (and is seen also in malnutrition).

Optic neuropathy with atrophy and central scotoma is relatively common in the alcoholic derelict.

Galen*:

Heavy wine, which heats up the intestines will increase the obstruction of the blood vessels and it will increase damage to the liver when inflammation and scirrhus already existed. After the liver, the spleen also can be damaged by sweet wine.

AMYLOIDOSIS

Suspect amyloidosis with:
1. periorbital purpura
2. macroglossia
3. refractory congestive heart failure
4. nephrotic syndrome
5. bilateral carpal tunnel syndrome
6. vitreous opacities
7. purple palmar creases, purple lid folds
8. orthostatic hypotension
9. hyaline plaques in skin folds (hyaline plaque deposition on the eyelids may be mistaken for xanthelasma)
10. history of chronic inflammatory disease (especially tuberculosis or rheumatoid arthritis)

* from RE Siegel, Galen's System of Physiology and Medicine, Karger, Basel, 1968.

ANGINA PECTORIS

Heberden (1772):

But there is a disorder of the breast marked with strong and peculiar symptoms. The seat of it, and sense of strangling, and anxiety with which it is attended, may make it not improperly be called angina pectoris. They who are afflicted with it are seized while walking (more especially if it be uphill, and soon after eating) with a painful and most disagreeable sensation in the breast, which seems as if it would extinguish life.... but the moment they stand still all this uneasiness vanishes.

Physical findings which may be appreciated during an episode are:

1. paradoxical splitting of P2 (indicating asynchronous ventricular contraction)
2. mitral systolic murmurs (of any type, secondary to papillary muscle dysfunction)
3. atrial diastolic gallop
4. transient pulsus alternans
5. cervical sympathetic activity (rare), left-sided pupillary dilatation.

When a patient presses a closed fist to his sternum in description of his pain, angina pectoris is likely to be present (Levine). Although the character of the pain varies considerably from patient to patient, in each case it is steady throughout its duration.

A rapid response to nitroglycerin is a valid indication that chest pain is angina. Acute attacks may at times be aborted with carotid sinus pressure (Levine). Treatment of intractable angina with continued carotid massage (i.e. electrical stimulation of carotid sinus nerves with an externally triggered, permanently implanted pacemaker) has been reported (NEJM 277:1278 '67).

Prinzmetal (variant) angina has the following features:

1. Age of onset is usually less than 50.

2. Although the pain has the character and typical distribution of angina (except that at times it persists somewhat longer) it may occur at rest, without obvious provocation.

3. Exercise tolerance may not be decreased in afflicted individuals; the exercise electrocardiogram will often not precipitate ST depression.

4. Attacks may be cyclical and of great regularity. In some individuals attacks are related to eating and can be precipitated with a test acid infusion through an N-G tube.

5. Pain is relieved with nitroglycerin and at times by exercise.

6. The electrocardiogram during an attack, reveals ST segment elevation.

The pathophysiology of Prinzmetal angina is not as yet known. In some cases reflex increased tone in a partially occluded coronary artery may be responsible. In a few patients with this disorder who have come to autopsy, complete occlusion of a major vessel has been present, indicating that the prognosis of this disorder is poor.

Preinfarction angina precedes 20 to 50% of cases of myocardial infarction. The anginal pain is typical for ischemia and perists anywhere from 15 minutes to several hours. Although diaphoresis may be present, there are no ST segment changes on EKG and no enzyme elevations occur. These patients should be managed with cardiac precautions as if infarction has occurred; anticoagulation with heparin appears to reduce the incidence of subsequent myocardial infarction significantly.

When chest pain suggests an initial diagnosis of pre-infarction angina, the occurrence of 4 or more coupled premature ventricular contractions (i. e. ventricular tachycardia) is prima facie evidence that infarction has actually occurred; other arrhythmias appear to be of no diagnostic significance (Wittenberg, NYU).

ANKYLOSING SPONDYLITIS

Ankylosing spondylitis occurs nine times more frequently in men than in women and generally presents during the young adults years. A genetic predisposition is suggested by twin concordance and family pattern studies.

Constitutional features are minimal or absent. LE preps are negative, and less than 5% of cases have rheumatoid factor. Anterior uveitis occurs in up to 50% of cases (at times antedating the arthritis) and may suggest this diagnosis when it occurs in a young man with sacroiliitis or lumbar muscle spasms. Prostatitis appears to be particularly common in men with this disease.

Aortic insufficiency is a rare concomitant and follows aortitis with dilatation of the valve ring.

Spotty upper lobe pulmonary lesions may be noted on the chest film, suggesting tuberculosis, although these changes represent a progressive fibrotic process (Brit J Dis Chest 59:90 '65). Chronic lung disease occurs late in the course of those with disabling ankylosis when vertebral distortion and chest wall restriction compromise pulmonary function.

THE AORTA

Dissecting aneurysm should be suspected when the clinical state appears to be shock, yet there is hypertension. Pulsation of the left sternoclavicular joint (sign of Logue and Sikes, JAMA 148:1209 '52) is a sign of this condition.

When the chest film following blunt chest trauma reveals widening of the superior mediastinum, traumatic aortic rupture is to be ruled out, regardless of the presenting state. Other X-ray changes are pleural effusion, loss of visibility of the aortic arch contour, and displacement of the left main stem bronchus. Apropos, aortic insufficiency after blunt trauma is more likely due to aortic rupture than to aortic cusp rupture.

Aortic coarctation is associated with intracranial aneurysm (and, therefore, subarachnoid hemorrhage). Three dimensional spiral tortuosity with 'serpentine' pulsations of the retinal arterioles are characteristic eye findings with coarctation (Trans Amer Oph Soc 50: 407 '52). (Retinal artery pulsation is always abnormal and occurs when intraocular and diastolic blood pressures are equalized.) Individuals with coarctation are often found to have bicuspid aortic valves.

Syphilitic aneurysms rarely dissect.

Roentgenographic signs of rupture of an abdominal aortic aneurysm on the plain abdominal film are (NEJM 265:12 '61):
1. (localized) loss of sharpness of the aneurysm margin with haziness, scalloping, nodularity, etc.
2. disruption of the calcific rim, extension of soft tissue density beyond the broken rim
3. loss of psoas margin
4. displacement of ureters (with IVP), (bowel is generally displaced anteriorly with rupture and on the lateral film, no gas overlies the soft tissue mass).
These findings are, of course, best appreciated in comparison with previous films when they are available.

Abdominal aortic aneurysms most often rupture into the retroperitoneal space. In the uncommon situation in which rupture is into the sigmoid mesocolon, physical examination will reveal striking perianal ecchymosis, a diagnostic and possibly pathognomonic finding. Other 'uncommon' sites of rupture with dramatic presentations are into the small intestine presenting as upper GI bleeding or into the vena cava presenting as acute congestive failure.

AORTIC VALVE DISEASE

In the presence of pulmonary emphysema, the murmur of aortic stenosis is often transmitted to the apex and may in fact be loudest in that location.

At least 75% of patients with pure aortic stenosis have congenitally deformed valves. When patients with severe (but apparently pure) aortic stenosis have a definite past history of rheumatic fever, concomitant, silent mitral stenosis should be suspected (Roberts, NIH).

An aortic ejection sound implies valvular mobility. It is present in virtually all cases of non-calcific aortic stenosis but less than half of those with extensive calcification.

Bicuspid aortic valves are not associated with hemodynamic abnormalities during youth but become symptomatic in later life when fibrosis or calcification occurs or when there is superimposed endocarditis.

Aortic stenosis is associated with occult upper GI bleeding, usually in older individuals. The site of bleeding is almost never located. The cause of this complication is unknown but may involve low output with ischemic necrosis of isolated mucosal regions.

Aortic insufficiency is associated with (Ann Int Med 47:27 '57) sudden death (10%) and with coarctation (3%). Patients may complain of:
1. atypical, angina pectoris (pain often nocturnal, may persist several minutes to an hour)
2. excessive sweating, heat intolerance (may suggest the diagnosis of hyperthyroidism)
3. throbbing neck pain
4. episodic abdominal pain (which may resemble that of peptic ulcer disease, mild pancreatitis, or cholelithiasis)
5. total body "pounding"
6. post-prandial "splashing"

When the murmur of aortic insufficiency is loudest to the right of the sternum, the lesion is very likely NOT rheumatic (Harvey, Georgetown).

Hill's test (comparison of brachial and femoral systolic blood pressure) is a valuable bedside means of estimating the severity of aortic insufficiency. If the difference is less than 20 mm Hg, AI is mild, from 20 to 40 mm Hg, AI is moderate, and when it is greater than 60 mm Hg, AI is gross.

ASTHMA

The development of asthma after the second decade without a positive family history requires consideration of the polyarteritis - Wegener's disease constellation.

Asthma can appear suddenly in adult Puerto Ricans with schistosomiasis.

"Monday morning asthma" is a feature of byssinosis. Tachyphylaxis is the mechanism. Monday morning headache in nitrate miners is another example of this sort of disorder.

When an acute asthmatic attack in children is associated with fever, pneumonia should be considered as a precipitating cause.

Inorganic arsenic may actually be therapeutic in cases of asthma refractory to conventional medicinals. It is used as K arsenite (Fowler's solution) in Gay's solution (which also contains digitalis, phenobarbital, and K iodide) (J All 40:327 '67).

'False positive' lung scans occur in asthmatics.

ATRIAL FIBRILLATION

The Gouaux-Ashman effect (Am H J 34:366 '47) is the occurrence of a transient bundle branch block terminating a short cycle following a long pause in a tracing with atrial fibrillation. This effect is an example of aberrated ventricular conduction which must be distinguished from ectopic ventricular activity. Aberrated conduction does not indicate digitalis toxicity, indeed, it may imply inadequate digitalization. Aberration is indicated when (Prog C V Dis 9:18 '66):

1. the anomalous beat in VI has a RBBB pattern with a triphasic (RSR') deflection (the RBBB pattern is of itself of no differential value)

2. the initial QRS vector of the anomalous beat is unchanged from previous 'normal' QRS deflections (a changed initial vector occurs with both aberration and ectopy).

Ectopy is favored when:

1. there is bigeminy or fixed coupling

2. the anomalous configuration terminates a long pause or occurs in an excessively pre-mature interval.

Aberration is also indicated when a therapeutic infusion of xylocaine increases rather than decreases the occurrence of anomalous beats.

Digitalis toxicity is suggested when a Wenckeback-Type EKG structure is observed, i.e. regular grouping of beats with progressive decrease in RR interval followed by a pause intermediate in length between the 2 preceding RR intervals.

Coarse atrial fibrillation (f waves greater than 1.0 mm in amplitude) indicates left atrial hypertrophy or strain (Circ 33:599 '66). When this finding is associated with a vertical mean QRS axis, mitral stenosis is a likely diagnosis, although with other axes no etiologic diagnosis is warranted. Fine atrial fibrillation, on the other hand, is more often an indication of atherosclerotic disease than of rheumatic or hypertensive cardiac abnormalities.

Atrial systole makes little hemodynamic contribution in the presence of significant mitral stenosis (Am J M 42:532 '67), although in other forms of heart disease, conversion of atrial fibrillation to sinus rhythm increases cardiac index and stroke volume.

ATRIAL MYXOMA

A left atrial myxoma can exactly mimic mitral stenosis. It should be suspected when:
1. the murmur and clinical status are associated with a change in position
2. signs develop suddenly in middle life
3. there is no history of rheumatic fever
4. the murmur of mitral stenosis is inconstant
5. cardiac decompensation is rapid and progressive, digitalis is ineffective

The physical examination discloses an "opening snap," which is (the so-called 'tumor plop') caused by the movement of the myxoma. Cine studies show that the tumor descends into the ventricle during early diastole and is rapidly forced back into the atrium during ventricular systole.

The laboratory examination will often reveal an elevated sedimentation rate and increased alpha-2 and gamma globulin levels. Embolic phenomena from the myxoma may initially suggest the diagnosis of subacute bacterial endocarditis, which is 'strengthened' by these laboratory findings.

Although 85% of myxomata occur on the left, in many cases they are multiple. Demonstration of a left atrial myxoma requires complete study of the right side of the heart as well.

Cardiac catheterization distinguishes myxoma from valvular stenosis by the absence of an early diastolic gradient. Time-motion ultrasoundcardiography is a hazard free means of making the diagnosis and has the advantage of being performed at the bedside with complete patient acceptability (Circ 39:615 '69).

BASAL CELL NEVUS SYNDROME

The characteristic facies of the basal cell nevus syndrome is one with widened inner canthal distance, supraorbital ridge prominence, and frontoparietal bossing.

The clinical features are (Ann Int Med 64:403 '66):

1. cutaneous - basal cell epitheliomata (appearing relatively early in life) and numerous, minute dyskeratoses of the palms and soles which slough readily, leaving the pathognomonic ICE PICK PALMS with hundreds of minute, red based pits.

2. osseous - short fourth metacarpal, spina bifida occulta (20% of cases), rib anomalies (particularly splaying), and bone cysts (particularly of the mandible).

3. genito-urinary - hypogonadism in males, ovarian fibromata in females.

4. miscellaneous - dura calcification, hyporesponsiveness to parathormone, autosomal dominant inheritance (with low penetrance).

BEHÇET'S SYNDROME

The development of pustular lesions with sterile venepuncture (or after sterile saline injection) is strongly suggestive of Behçet's syndrome.

Recurrent oral and genital mucosal ulcerations are the hallmark of this disorder (and also of pemphigus vulgaris which must be excluded in the differential diagnosis). Other, variable, features are:

1. ocular lesions - severe uveitis with hypopyon which may simulate endophthalmitis

2. CNS dysfunction (at any level) - CSF pleocytosis is usual and the CSF protein is above 50 mg% in more than half of cases (Am J Med 34:544 '63)

3. erythema nodosum

Inspection of the tongue with a magnifying lens reveals absence of the fungiform papillae from the anterior region and proliferation of the filiform papillae, a definite sign of this syndrome (Arch Int Med 124:720 '69).

Thrombophlebitis is a frequent concomitant of this syndrome, although no definite coagulation disorder has as yet been demonstrated.

BOTULISM

Muscle weakness is responsible for most clinical signs during the early stages of intoxication. Paresis of accommodation with dimming of vision is usually first noted, ptosis and extraocular muscle paralyses follow. Dysphagia and dysphonia are later developments. Sensory changes are not present, and the deep tendon reflexes are preserved.

Mydriasis is an early feature of botulism. Clinical distinction from belladonna intoxication is possible since in botulism mental status is normal until the final stages.

Guanidine may be used to treat the neuro-muscular effects (NEJM 278:931 '68).

BRUCELLOSIS

Brucellosis may present either as a chronic process resembling tuberculosis or as an acute febrile disease. Contact with domestic animals (or their products) is an essential historical feature. This diagnosis should be considered when a person from a rural area:

1. has a febrile disorder which improves with bed rest and promptly recurs with activity

2. is thought to have infectious mononucleosis but confirmatory findings are lacking

3. has a febrile disorder and complains of pain on motion of the eyes (inflammation of Tenon's capsule is characteristic of this disease)

4. has marked weakness, easy fatigueability, anorexia, and night sweats but pulmonary tuberculosis cannot be confirmed

Emotional disturbance, namely, psychoneurosis, anxiety, irritability, and/or depression is a constant feature of the chronic type of the disease. A brucella skin test should be included in the initial evaluation of 'rustics' who have had clear cut personality changes.

BURNS

Death following major burns may be due to:
1. damage to the respiratory tract (major cause!)
2. shock, electrolyte imbalance
3. renal failure (acute tubular necrosis)
4. heart failure, arrhythmias
5. activation of an underlying systemic disorder (such as a collagen vascular disease)
6. sepsis
7. pulmonary embolus
8. adrenal hemorrhage
9. GI hemorrhage (Curling's ulcer)
10. drug reactions

Microangiopathic hemolytic anemia and thrombocytopenia often follow major burns and may be delayed and persistent.

Pseudomonas superinfection can be detected by repeated examination of the burn site for fluorescence with a Woods' light (JAMA 202: 1039 '67).

CANCER

Cancer may present with:
1. thrombophlebitis (Trosseau's syndrome)
2. endocrine abnormalities
3. cerebellar ataxia (particularly lung CA)
4. refractory neuropsychiatric depression (pancreas)
5. myasthenic symptoms (see "MYASTHENIA")
6. epidermoid cysts, osteomata (Gardner's syndrome)
7. pruritis, skin changes (see "SKIN")
8. polymyositis, dermatomyositis
9. digital ischemia (without prior Raynaud's phenom., Brit M J 3:208 '67), clubbing
10. umbilical nodule (Sister Joseph's nodule)
11. thrombocytosis, anemia, white cell changes (see below)

46

12. hemorrhagic diathesis, hypofibrinogenemia
13. non-bacterial (marantic) endocarditis (widespread cancers associated with derangement of intravascular clotting such as pancreas, stomach, and lung adeno- CA).

Human tumors are weakly antigenic; an autoimmune response to some tumors can (rarely) be detected (B J Ca 23:510 '69). In the future it may be possible to treat neoplastic disease by inducing endogenous lymphocytes to produce specific antibodies.

Anergy to skin testing with 2, 4 dinitrochlorobenzene (DNCB) implies a very poor prognosis, regardless of the histologic type of tumor, its anatomical location, or its apparent clinical extent (Surg Forum 20:116 '69).

'Malignancy associated changes' (MAC) in the cytological features of a variety of cells have been noted, indicating a systemic cellular concomitant of cancer (Acta Cytol 11:415 '67). Indeed, greater than 90% accuracy in identifying cases of cancer by screening for changes in the nuclei of polys and the cytoplasm of monos of venous blood has been reported (Acta Cytol 12:313 '68). Electron microscopy of polys may eventually disclose certain malignancy 'specific' features (Acta Cytol 13:435 '69).

CARCINOID SYNDROME

Patients with asthma (or chronic obstructive lung disease) may mistakenly be thought to have carcinoid syndrome if it is not recalled that glyceryl-guaiacolate (the active agent of many expectorants) gives a false positive 5HIAA test.

The flush of the intestinal carcinoid is evanescent and involves the "blush" area, which in time takes on a permanent telangiectatic change. The flush of a bronchial carcinoid is distinctive being prolonged, extensive, florid, and associated with conjunctival injection, lacrimation, excessive salivation, tachycardia, and, at times, oliguria. Eventually the skin of the face becomes thickened, the facies leonine. Because of salivary gland enlargement, there is drooling or dribbling. The eyes are characteristically wet and shiny. The flush of the gastric carcinoid is also distinctive, being a blotchy, evanescent, raised, indurated, and quite pruritic eruption involving the face, trunk, back, and (less often) palms.

In general, whatever causes one to blush will precipitate a carcinoid flush. Alcohol induces flushing readily. Flushing in bronchial carcinoid may be life endangering because of marked hypotension. It is at times controlled by phenothiazine use or, when this fails, by steroids.

The eye signs of carcinoid are (NEJM 277:406 '67):
1. macular pigment clumping
2. scattered pigmentary specks overlying venules
3. exudative spots, which resolve to punched out scars with a pigmented border.

The gastric carcinoid differs chemically from intestinal and bronchial carcinoids in that 5 hydroxytryptophan rather than serotonin is released.

The cardiac complication of carcinoid is the deposition of plaque-like material on the endocardium of the right side of the heart, particularly on the 'arterial' sides of the pulmonic and tricuspid

valves and the chordae of the tricuspid valve, leading to tricuspid insufficiency and pulmonic stenosis. Cardiac changes follow metastases to
the liver and are <u>almost</u> always associated with the complete syndrome.
Plaque deposition may also occur in the left side, particularly when
there is an intraatrial defect and reversed flow or when there is a bronchial carcinoid.

CARPAL TUNNEL SYNDROME

Bilateral carpal tunnel syndrome follows occupational, excessive,
hand use for the most part, but it can occur with the following systemic
conditions:

1. diabetes
2. amyloidosis
3. sarcoidosis
4. acromegaly
5. hypothyroidism
6. Cushing's disease
7. chronic tophaceous gout (possibly latent gout as well)
8. (transiently) pregnancy

Carpal tunnel syndrome often begins insidiously as nocturnal
acral paresthesias. Physical exam may reveal thenar wasting or ting-
ling in a median distribution in the hand with light percussion over the
median nerve at the wrist (Tinel's sign). The tourniquet test, however,
(Lancet 2:595 '53) is the most useful (simple) diagnostic test (ischemia
is induced by inflating a cuff above the elbow; with carpal tunnel syndrome paresthesias in the median distribution are elicited; with very
early compression changes, there are no paresthesias, but median distribution sensory loss occurs within 10 minutes).

THE CHINESE RESTAURANT SYNDROME

The CHINESE RESTAURANT SYNDROME (Sci 163:826 '69) occurs when susceptible individuals eat foods prepared with mono-sodium
l-glutamate, as, for example, Wonton Soup in Chinese restaurants or
meats seasoned with "Accent" at home. The symptoms are:

1. burning paresthesias, usually of the face, neck, or arms,
varying in intensity from mild prickling to severe discomfort

2. facial "pressure" or numbness, typically of the malar emi-
nences

3. chest pain resembling angina (but without associated EKG
changes)

4. facial flushing, facial diaphoresis

5. lightheadedness

The severity and number of different symptomatic complaints
appears to be dose related, although the threshold of sensitivity varies
with individual idiosyncrasy and with the amount of food already in the
stomach before exposure.

COLITIC ARTHRITIS

Colitic arthritis is not a form of rheumatoid arthritis. RA factor and LE preps are negative, men and women are equally afflicted,

48

peak age of incidence is during the 3rd decade. It is seen in 75% of patients with Whipple's disease, 15% of those with ulcerative colitis, and 5% of those with Crohn's disease (intestinal or colonic).

The arthritis is:
1. monarticular (asymmetric)
2. migratory
3. large joint (knees and ankles most often)
4. non-deforming
5. without X-ray changes

Colitic arthritis may be a presenting feature or may follow intestinal symptomatology. It abates with surgical removal of the involved segment of gut.

25% of individuals with colitic arthritis have sacroiliitis, other concomitants are:
1. ocular involvement, usually episcleritis, 15+% of cases (somewhat less with Crohn's disease)
2. erythema nodosum, 20%
3. pyoderma gangrenosum, 5% (usually ulcerative colitis)

CONGENITAL HEART DISEASE   (See also "TETRALOGY OF FALLOT")

Sweating is increased in children with congenital heart disease who are prone to congestive failure, regardless of the anatomical defect or the state of compensation when sweating is evaluated. The pilocarpine sweat test may be a useful prognostic indicator (Ped 41:123 '68).

When pulmonic stenosis is associated with atrial septal defect, the stenosis is almost always valvular and the septal defect is of the os secundum type. When VSD is associated with pulmonic stenosis, the stenosis is infundibular more often than valvular. An ejection click localizes the stenosis to the valvular level (Perloff, Georgetown).

When the clinical signs suggest ASD, but the EKG reveals a QVI pattern (mimicking an intracavitary tracing and indicating massive right atrial enlargement), consider anomalous pulmonary venous drainage in the differential diagnosis. When ASD is associated with a QV6 pattern, the defect is likely of the os primum type.

Left axis deviation occurs in less than 5% of uncomplicated VSD's. When this seems to be the case, double outlet right ventricle and endocardial cushion defect should be considered in the differential diagnosis.

When the history and murmur are indicative of a ventricular septal defect but other findings are not in accord, corrected transposition of the great vessels should be considered. In this disorder the 'anatomical' and 'physiological' ventricles are reversed, the aorta is displaced anteriorly, the pulmonary vessels shifted to the right, and the interventricular septum oriented perpendicular to the chest wall. The correlates of these changes are:

1. the arterial ventricular impulse extends from the apex to the sternal edge, suggesting RVH

2. there is a loud aortic closure sound in the pulmonic area which may be mistaken for a sign of pulmonary hypertension

3. the chest film reveals a narrow vascular pedicle; the pulmonary vasculature makes no contribution to the left heart border; the

infundibulum of the (anatomic) right ventricle makes a characteristic hump along the left heart border.

4. electrical depolarization proceeds from right to left (the septum is anatomically reversed); right precordial and inferior limb lead Q waves are present; A-V conduction defects are common

Surgically difficult to repair, complex, cyanotic congenital cardiac defects occur with ASPLENIA which may be suspected when the cyanotic infant has hematologic changes indicating the functional absence of this organ, namely, nucleated RBC, H-J bodies, siderocytes, and target cells on peripheral smear and decreased RBC osmotic fragility. The X-ray finding of a transverse lower hepatic margin is confirmatory. The cardiovascular defects are some combination of (Am J Cardiol 13:386 '64):
1. ASD, VSD, or single ventricle
2. common A-V canal
3. transposition of the great vessels
4. infundibular or ventricular inversion
5. pulmonic stenosis or atresia
6. anomalous pulmonary or systemic veins

When there are left to right intracardiac shunts, strangely enough there appears to be an inverse relation between the degree of pulmonary hypertension and the dermatoglyphic feature, the total finger ridge count (Acta Cardiol 24:382 '69). When the total ridge count is greater than 200, pulmonary hypertension is minimal if at all present; when the count is less than 50, pulmonary hypertension is moderate to severe.

CUSHING'S SYNDROME

Cushing (Bull JHH 50:137 '32):

Physical examination. This showed an undersized, kyphotic young woman 4 feet 9 inches in height (145 cm), of the most extraordinary appearance. Her round face was dusky and cyanosed and there was an abnormal growth of hair, particularly noticeable on the sides of the forehead, upper lips and chin. The mucous membranes were of bright color despite her history of frequent bleedings. Her abdominous body had the appearance of a full term pregnancy. The breasts were hypertrophic and pendulous and there were pads of fat over the supraclavicular and posterior cervical regions. The cyanotic appearance of the skin was particularly apparent over the body and lower extremities which were spotted by subcutaneous ecchymoses. Numerous purplish striae were present over the stretched skin of the lower abdomen and also over the shoulders, breasts and hips; and a fine hirsuties was present over the back, hips, and around the umbilicus. The skin which was everywhere rough and dry showed considerable pigmentation, particularly around the eyelids, groins, pubes, and the areolae of the breasts. The peculiar tense and painful adiposity affecting face, neck, and trunk was in marked contrast to her comparatively spare extremities.

(from Ross, Marshall-Jones, and Friedman, Q J Med 35:149, 193 '66):

All published series show a predominance of females. Cushing's syndrome in the male is rare. On the other hand, adrenocortical hyperplasia secondary to extra-adrenal carcinoma is commoner in males than in females. When a male is seen with the clinical features of Cushing's syndrome, careful search should be made for the possible presence of an underlying carcinoma.

Most obese, red-faced hirsute women do not suffer from adrenocortical overactivity. Except in florid cases, a definite diagnosis is not easily made on clinical grounds.

Adrenocortical overactivity should be included in the differential diagnosis of all patients exhibiting any one of the following features:

hypertension, particularly in young people;

glycosuria, especially if the fasting blood sugar is normal;

osteoporosis, particularly in patients who have not reached the menopause, in patients who fracture bone with trivial trauma; in patient's whose fractures do not heal or which heal with excessive formation of callus;

in patients with renal calculi or nephrocalcinosis;

in patients with hypercalciuria without hypercalcemia;

in patients with amenorrhea or irregular periods;

in adolescent patients who cease to grow; and

in patients with psychological difficulties.

Plasma electrolyte concentrations are of little assistance in the diagnosis of the syndrome, with the important exception that hypokalemic alkalosis is a common complication of bilateral adrenocortical hyperplasia secondary to a cancer of other organs and also of adrenocortical carcinoma..... In contrast with the classical non-malignant Cushing's syndrome, hypokalemic alkalosis is almost the rule in Cushing's syndrome associated with malignancy.

DIABETES MELLITUS

With diabetes of sudden onset, the following possibilities should be considered:

1. pheochromocytoma
2. Cushing's disease
3. hyperthyroidism
4. pituitary tumors
5. pancreatitis, pancreatic neoplasm
6. thiazide use

Calcification of the vas deferens may be a pathognomonic X-ray finding (Roent Clin NA 5:239 '67).

The skin changes of diabetes are:

1. diabetic dermopathy (Arch Derm 92:625 '65) - 5 to 10 mm multiple, indurated, flat, dull red papules of the extensor extremity surfaces. A central cell may be present. With time, these lesions become atrophic and hyperpigmented

2. diabetic shin spots - 5 to 25 mm hyperpigmented macules of the pretibial zone. These occur in about 50% of juvenile and adult onset types but in less than 2% of the non-diabetic population (Am J Med Sci 251:570 '66).

3. pruritis

4. necrobiosis lipoidica

5. xanthoma diabeticorum

6. pyogenic or monilial superinfection

7. bronzing (when hemochromatosis is causative)

8. keratinization and yellowing of the palms and soles (secondary to carotenemia)

Concomitant premature vascular calcification and osteoporosis should suggest diabetes. Some degree of osteoporosis is noted in about half of adult onset type cases and appears to involve the thoracic and lumbar vertebral bodies most noticeably early in the course.

Femoral nerve palsy is the most frequent peripheral nerve defect in diabetes. Conversely, proximal muscle weakness of the legs or an isolated quadraceps femoris paralysis should suggest diabetes. A third nerve palsy, with sparing of the pupillary reactions, is the commonest cranial nerve defect in diabetes.

Autonomic neuropathy is indicated by:
1. orthostatic hypotension
2. impotence
3. altered sweating (particularly localized anhidrosis)
4. altered bowel activity (constipation is most frequent, diarrhea less often, and nocturnal diarrhea least common and most specific)

Orbital cellulitis suggests mucormycosis, especially in the presence of acidosis. Nasal scrapings may provide organisms for culture and microscopic examination. Rapid treatment is imperative. (Apropos, orbital mucormycosis can occur with any debilitating condition predisposing to acidosis).

The eye signs of diabetes are:

1. frequent changes in refractive error (due to changing states of lens hydration)

2. cataract (usually non-specific)

3. rubeosis iridis (neovascular formation of the iris, later leading to glaucoma)

4. retinopathy (characteristically - microaneurysms, hemorrhages and exudates scattered about the posterior pole; may progress to a proliferative retinopathy with neovascularization, vitreous hemorrhage, and retinal detachment

5. lipemia retinalis (transient, occurring with blood lipid in excess of 1200 mg%)

6. IIIrd nerve palsy (with pupillary sparing), (less often) IVth nerve palsy

Retinopathy is present in about 20% of all diabetics (juvenile and adult onset). Of those with retinopathy, at least 80% have diabetic nephropathy; when retinitis proliferans is present, renal involvement is certain and life expectancy is limited.

Central retinal vein occlusion should lead to an investigation for diabetes if the diagnosis has not previously been made.

Crepitant wound infections generally imply a Clostridial infection, but in diabetics this finding may occur with E. Coli. In diabetics, tuberculosis may involve 'atypical' regions, such as the lower lobes.

BACTERIAL ENDOCARDITIS

Histiocytosis in the smear of the first drop of blood obtained from an ear lobe strongly suggests the diagnosis of subacute bacterial endocarditis, although the absence of this finding does not militate against this diagnosis (J Lab Clin Med 48:827 '56). The presence of actively phagocytic histiocytes may perhaps be pathognomonic in a patient with an unidentified chronic disease process (Am J Med 26:965 '59).

Although bacterial endocarditis is a frequent complication of ventricular septal defect, its occurrence with an atrial septal defect is extremely rare.

Pneumococcal endocarditis localizes to the aortic valve. Streptococcal endocarditis can occur on a 'normal' valve, particularly in the elderly and should be considered in the differential diagnosis of stroke, anemia, and unexplained fever.

Sternal tenderness is common.

Purpura without abnormalities of platelet morphology, clot retraction, bleeding time, clotting time, or tourniquet test may be a presentation of subacute bacterial endocarditis (Silber, NYU). This sign may also be noted with typhoid, chronic meningococcemia, rickettsial diseases, dysgammaglobulinemias, and Henoch-Schönlein disease.

Bacterial endarteritis developing in a thrombosed aneurysm or at the site of an A-V fistula should be considered when the clinical picture suggests subacute bacterial endocarditis but the cardiac examination is without abnormality.

Right sided endocarditis involving the tricuspid valve is usually clinically 'silent', and in this condition blood cultures are most often negative. It is usually seen in drug addicts with repeated, unsterile intravenous instrumentation, and the infecting organism is usually a staphylococcus or candida leading to sepsis with multiple suppurative abscesses, particularly of the lung.

FABRY'S DISEASE
(angiokeratoma corporis diffusum universale
Am J Med 42:293 '67)

Fabry's disease is a systemic disorder with manifestations consequent to the generalized deposition of a glycolipid ceramide trihexose, a normal component of the red cell stroma. Inheritance of this disease follows a sex-linked recessive pattern, although there is occasionally penetrance in heterozygous women. Fabry's disease occurs rarely, if ever, in Negroes.

Ceramide trihexosidase activity is extremely low in small intestine biopsy specimens from homozygous patients, suggesting that deficiency of this enzyme is the underlying cause of the disease (Laster, NIH).

Suspect this disease when there are:
1. angiokeratomata of the skin, especially in the region of the umbilicus or of the scrotum
2. severe acral paresthesias
3. unexplained dependent edema
4. heat intolerance secondary to an- or hypo-hidrosis

The usual features are:

skin - purple polygonal patches in a bathing suit distribution or on the oral mucosa. Skin lesions appear during childhood and are associated with acral pains and episodic fever which intensifies with time. Acral pains are often precipitated by exercise or local heat.

kidney - asymptomatic proteinuria with lipoiduria and foam cells occurs during the 2nd decade and is followed by progressive renal decompensation during the next twenty years.

eyes - posterior 'spider-like' lens opacification (characteristic), bronze, whorl opacities of the cornea (found only with Fabry's disease and chloroquine toxicity), episcleral vessel changes (which may be detected with the hand ophthalmoscope, see "CONJUNCTIVA"). A careful slit-lamp examination will reveal an infiltrate around involved episcleral vessels, biopsy in this region (an office procedure) will give the definitive diagnosis.

heart - glycolipid deposition within the myocardium leads to cardiomegaly and subsequent cardiac failure; hypertension (about a third of cases).

FAMILIAL MEDITERRANEAN FEVER

Familial Mediterranean Fever is an autosomal recessive trait with decreased penetrance in females.

Its incidence is about 1:500 Armenians and 1:2000 Sephardic Jews (i.e. those of Spanish, Mediterranean, and North African ancestry). There are approximately 80 new cases each year in New York City (Pras, NYU).

The clinical features are:

1. acute peritonitis (95%) - sudden onset of severe abdominal pain in association with high fever. Peritoneal signs are present on examination. Symptoms generally subside within 24 hours and recur at irregular intervals.

2. acute synovitis (75%) - typically there is monarticular involvement of the lower extremity large joints. Although involved joints are quite swollen, there is relatively little heat or erythema. Joint symptoms generally persist for 2 or 3 days, though after many attacks, a chronic arthritis occurs.

3. acute pleuritis (40%)

In the interval between attacks, patients are generally in good health, which helps to distinguish this disorder from other causes of periodic fever. In many instances, there is remission during pregnancy. Eventually amyloidosis occurs, presenting, generally, as nephrosis.

Laboratory testing may reveal increased fibrinogen, increased alpha $_2$ globulin, and increased glycoprotein.

GASTRIC CARCINOMA - GASTRIC ULCER DISEASE

A preferential LOSS OF TASTE FOR MEAT is seen in most cases of gastric carcinoma.

The symptoms of a benign gastric ulcer should abate within a few days of initiating intensive medical therapy; complete healing should be present roentgenographically within 2 months. When these conditions are not met, surgery is indicated, for either there is cancer or the ulcer is intractable and may perforate.

The morphine-primed cine gastrogram may aid in the differential of gastric ulcer and gastric neoplasm (Donner, JHH). Morphine induces a strong contraction wave which does not pass smoothly through a region infiltrated with carcinoma.

When free blood is present in the stomach, free acid will not be detected. However, when clotted blood is expelled or aspirated from the stomach, achlorhydria is likely.

GONOCOCCAL ARTHRITIS

Gonococcal arthritis occurs more frequently in women than in men and appears to occur in two forms:

1. (the classical) monarticular, septic form without systemic symptoms, and

2. a polyarthritic, non-septic form with definite systemic manifestations.

The polyarthritic form occurs four to five times more often than the monarticular form. Joint involvement occurs in the following order of frequency: knees, wrists, ankles, elbows. The features are:

1. systemic
   a. fever, 100%
   b. chills, 70%
   c. skin manifestations, 50%

2. extra-articular
   a. tenosynovitis, 50% (may be difficult to evaluate in the presence of an acute arthritis)

3. laboratory
   a. leukocytosis, 90% (of itself of no diagnostic value)
   b. + blood cultures (definitive when present)
   c. inflammatory joint fluid which is sterile!
   d. X-rays without septic joint changes (despite the historical length of the process)

Response to penicillin is of little diagnostic significance, although it is certainly clinically gratifying. Gonococcal polyarthritis may initially be confused with Reiter's syndrome (in men) and with a collagen vascular disorder (in women), particularly when it is recurrent and associated with skin changes.

GOUT

Sydenham (1683):

As often as gout is regular, it comes on thus. Towards the end of January or the beginning of February, suddenly and without premonitory feelings, the disease breaks out. Its only forerunner is indigestion and crudity of the stomach, of which the patient labours some weeks before. His body feels swollen, heavy, and windy - symptoms which increase until the fit breaks out. This is preceeded by a few days of torpor and a feeling of flatus along the legs and thighs. Besides this, there is a spasmodic affection, whilst the day before the fit the appetite is unnaturally hearty. The victim goes to bed and sleeps in good health. About two o'clock in the morning he is awakened by a severe pain in the great toe; more rarely in the heel, ankle, or instep. The pain is like that of dislocation, and yet the parts feel as if cold water were poured over them. Then follow chills and shivers, and a little fever. The pain which was at first moderate becomes more intense. With its intensity the chills and fever increase. After a time this comes to its height, accommodating itself to the bones and ligaments of the tarsus and metatarsus. Now it is a violent stretching and tearing of the ligaments - now it is a gnawing pain and now a pressure and tightening. So exquisite and lively meanwhile is the feeling of the part affected, that it cannot bear the weight of bedclothes nor the jar of a person walking in the room. (Latham translation).

Osler (1914): (The Principles and Practice of Medicine, 8th ed.
              D. Appleton and Co.)

The gouty polyarthritis may be afebrile. A patient with three of four joints red, swollen, and painful in rheumatic fever has pyrexia, and, while it may be present and often is in gout, its absence is, I think, a valuable diagnostic sign.

Patients with chronic gout are usually dyspeptic, often of a sallow complexion, and show signs of arteriosclerosis.

Alcohol is an important factor in the etiology. Fermented liquors are more apt to cause it than distilled spirits, and the disease is much more common in England and Germany, the countries which consume the largest amount of beer per capita.

Chronic bronchitis occurs with great frequency in persons of a gouty habit.

Premonitory symptoms (of acute gout) are common - twinges of pain in the small joints of the hands and feet, nocturnal restlessness, irritability of temper, and dyspepsia.

A common gouty manifestation, upon which Duckworth has laid stress, is the occurrence of hot or itching feet at night..... Cramps in the legs may also be very troublesome. Hutchinson has called attention to hot and itching eyeballs. Associated with or alternating with this symptom there may be attacks of episcleral congestion.

PROSTHETIC HEART VALVE

Turbulent flow, particularly across a prosthetic aortic valve, may lead to red cell fragmentation with typical microangiopathic hemolytic anemia. Similar changes can also occur with severe rheumatic valvular distortion. It is advisable to test for red cell fragility (with the milipore technique) prior to valvular replacement.

Infection can gain entry via a respiratory tract infection or can be introduced at the time of surgery despite adequate antibiotic coverage. In the later case, clinical presentation usually occurs within 12 weeks, although it may be delayed 3 to 6 months when coagulase negative staph is the invading organism.

1. 30% of cases present with suture separation, indicated by the sudden appearance of a regurgitant murmur,

2. slightly less than 30% have embolic phenomena,

3. about 10% have thrombus formation.

Coagulase negative staph are involved in about 30% of cases, coagulase positive 25%, unspecified staph 10% (JHH).

Infection is to be differentiated from
1. post-perfusion syndrome
2. post-pericardiotomy syndrome
3. drug reaction
4. sternal osteomyelitis

HEMOCHROMATOSIS

Clinical hemochromatosis occurs 10 to 20 times more often in men than in women, probably because menstruation is therapeutic in this instance. Although the genetic defect of primary hemochromatosis is as yet uncertain, it is wise to determine the plasma iron level of all adult relatives of patients so that the disorder may be detected in the preclinical state.

Iron stains of venous blood - buffy coat smears will reveal iron containing macrophages (on the order of 1 to 500 white cells) in patients with this disorder prior to therapy (NEJM 279:512 '68). This procedure may be of some value at the time of presentation as a screening test when suspicion of hemochromatosis is present.

Chondrocalcinosis and calcification of the menisci of the knees and the ligaments of the wrist are frequently associated with hemochromatosis. In about 50% of cases there is a characteristic arthropathy (Q J Med 37:171 '68) which involves the small joints of the hands, the wrists, ankles, knees, and hips. X-rays demonstrate:

1. osteoporosis of the phalanges and metacarpals
2. juxta-articular cyst formation, especially of the wrist
3. narrowing of the joint spaces with some degree of fragmentation of the articular surfaces

HEMOGLOBINOPATHIES (See also "SICKLE CELL ANEMIA")

Microscopic hematuria is a frequent finding in individuals with sickle cell trait and a common observation in patients with S-C disease. In both instances bleeding is from one kidney, with a left renal source slightly more common than the right. About 70% of individuals with sickle trait have hyposthenuria.

Homozygous C disease is clinically benign. Episodic arthralgias may occur. The peripheral smear reveals target cells and microspherocytes. Crystallization occurs when homozygous C red cells are suspended in hypertonic saline, an unique property of this type of hemoglobin.

Women with S-C disease are prone to obstetric complications, namely increased anemia during pregnancy and crisis near or following delivery (the mechanism of this complication may involve marrow and fat embolization).

Some 86% of patients with S-C disease have eye signs (which can be detected with the hand ophthalmoscope during the admission examination) (Arch Ophth 75:353 '66). These are:

1. the Paton curleyque conjunctival vessel sign, 80 +% (see "CONJUNCTIVA")
2. obliteration of peripheral arterioles and venules, 73%
3. A-V sea fans, 59% (practically pathognomonic, although it is, rarely, also, seen with S-Thal. One case of chronic myelogenous leukemia with hyperviscosity syndrome (AA_2 hemoglobin) and seafans has been seen at the JHH.
4. refractile lipid exudates, 36%
5. retinal vascular proliferation, 20%
6. "sun burst" pigmentation, 18%

HEPATIC DISEASE

A right upper quadrant mass WITH A PALPABLE EDGE must be liver.

Decreased renal blood flow occurs with tense ascites, further accentuating fluid retention. Paracentesis may be helpful in this situation.

Hepatic decompensation with sudden personality change implies hepatoma and is a very poor prognostic sign (Tillett's sign).

Bradycardia accompanies deep jaundice and may represent bile salt depression of the SA node (Austin Flint).

Postanesthesia jaundice carries a poor prognosis and occurs with ALL anesthetics, not just halothane. Hepatic damage appears to be on an allergic basis. Fever is a prominent clinical feature; jaundice is first noted about 1 week post-op and peaks 1-2 weeks later. Repeated liver challenge significantly increases the chance of severe liver damage, making early recognition of hepatic insufficiency mandatory before reoperation.

Osler:

A not infrequent situation (of an hydatid cyst) is to the left of the suspensory ligament, the resulting tumor pushing up the heart and causing an extensive area of dullness in the lower sternal and left hypochondriac regions. In the right lobe, if the tumor is on the posterior surface, the enlargement of the organ is chiefly upward into the pleura and the vertical area of dullness in the posterior axillary line is increased.

... frequent is the mistake of confounding a hydatid cyst of the right lobe pushing up the pleura with pleural effusion of the right side... Cysts of moderate size may exist without producing symptoms. Large multiple echinococci may cause great enlargement with irregularity of the outline, and such a condition persisting for any time with retention of health and strength suggests hydatid disease.

With jaundice, although "sweat tinges the linen" (Osler), tears, saliva, and milk are rarely stained.

The combination of cerebellar and basal ganglion symptoms suggests Wilson's disease but can also be seen in chronic hepatic disease with portal-systemic shunting. Patients in this later category may also complain of heaviness or clumsiness of the lower legs or even paraplegia, which is secondary to irreversible pyramidal tract involvement (see Q J Med 36:135 '67).

Subclinical steatorrhea is a frequent laboratory finding in alcoholics with Laennec's cirrhosis. In some instances this is reversed with pancreatic replacement therapy (Arch Int Med 119:567 '67).

Serum fetal antigens appear with some neoplastic disorders. "Alpha₁ fetoprotein" is present in slightly less than half of cases of primary hepatic carcinoma and may well prove to be diagnostic of this condition (Lancet 2:1267 '69).

HOMOCYSTINURIA (Laster, NIH)

Invariable clinical features:
1. mental retardation
2. fine, light hair; fair complexion
3. malar flush
4. peculiar gait, knock knees
5. ectopia lentis (see "LENS") (invariably by age 10)
6. fatty liver

Variable features:
1. pes cavus
2. buck teeth
3. marfanoid extremities and phalanges
4. cardiovascular disorders
5. thrombotic incidents
6. convulsions
7. gluten sensitive enteropathy

The basic biochemical defect is the absence of cystathione synthetase activity. As a result, cysteine cannot be synthesized and serum homocystine and methionine levels are increased. Although cysteine is present in the adult diet, its synthesis from methionine appears to be essential in infants for proper cerebral development.

Cystathione synthetase activity is partially reduced in those few asymptomatic relatives of patients who have been studied to date, suggesting a recessive genetic mechanism.

Therapy should include cysteine supplementation as well as pyridoxal phosphate, a synthetase cofactor, which in some instances reduces or removes homocystine from the urine.

Fine, blond hair, fair complexion, and mental retardation (without the other features of homocystinuria) are also seen in phenylketonuria and in the rare disorder, the neural crest syndrome (Arch Neurol 15:294 '66). Other features of this latter condition are:

1. loss of superficial and/or deep sensation
2. anhydrosis, pupillary abnormalities, vasomotor instability (autonomic effects)
3. dental enamel aplasia (yellow teeth)

4. hyporeflexia
5. meningeal thickening

HORNER'S SYNDROME

The features of the complete syndrome are unilateral miosis, ptosis, enophthalmos, and facial anhydrosis.

In a review of 216 cases (Am J Ophth 46:289 '58), causes were:
1. neoplasia 36% (malignant:benign = 3.1)
2. surgical 18%
3. trauma (including birth trauma) 13%
4. vascular 4%
5. miscellaneous 4%
6. "idiopathic" 25%

The terminal group of sympathetic fibers for facial sweating course with the external carotid. Therefore, when anhydrosis is NOT a feature, the causal lesion must be located distal to the carotid bifurcation.

Miosis, ptosis, enophthalmos (apparent, due to ptosis), and severe headache in the trigeminal distribution, all without anhydrosis is known as Raeder's syndrome and is caused by a (benign) para-trigeminal lesion.

The commonest intracranial cause of Horner's syndrome is Wallenberg's syndrome. In the admittedly uncommon situation of a central Horner's syndrome, topical cocaine causes dilatation of the involved pupil, a specific, 'localizing' test.

If the sympathetic lesion can be localized to the sympathetic trunk, and surgery or trauma are excluded by history and physical, NEOPLASIA IS RESPONSIBLE FOR MORE THAN 90% OF CASES, and of these, malignancy is three times more common than a benign tumor. Apropos, the 'pathognomonic' X-ray picture of a Pancoast tumor is an apical lung opacification with erosion of the inner surfaces of the upper ribs.

A partial form of Horner's syndrome, namely minimal ptosis and miosis which is apparent in dim light (though not in bright light since pupillary reactions are preserved) occurs with some cases of acute, purulent otitis media and represents involvement of sympathetic fibers from the carotid plexus as they pass through the middle ear (NEJM 265:475 '61). Anhydrosis is not a component of this constellation.

Dynamic exophthalmos (protrusion of the globe with ocular motion), mydriasis, and slight lid retraction is Claude Bernard's syndrome and represents irritation of the cervical sympathetic tract. It is seen with inflammatory or infiltrative processes involving the cervical sympathetic chain and:

1. rabies
2. lead poisoning

HYPERTENSION

Sudden development and continued progression of hypertension (or its sudden acceleration, especially beyond the 4th decade), with normal BUN, urinalysis, and hematocrit, raises the suspicion of:

1. adrenal tumor
2. collagen vascular disease
3. bacterial endocarditis
4. renoprival hypertension

When hypertension is associated with a recent history of abdominal pain, dissecting aneurysm with obstruction of renal blood flow should be considered.

Hypertension and hypokalemia:
1. primary or secondary hyperaldosteronism
2. chronic licorice ingestion (the active agent is glycyrrhizic acid, a potent sodium retainer - potassium diuretic)

Foods high in tyramine content are cheeses, herring, chopped chicken liver, and chianti. Hypertensive crisis may occur when individuals on MAO inhibitors indulge in these items.

The ocular fundus of individuals with hypertension of short duration without prior vascular disease exhibits cytoid bodies and retinal edema (the characteristic wet, shiny, "shot silk" fundus) with arterioles of normal caliber. Eclampsia, acute renal failure, and pheochromocytoma are the most frequent of such conditions.

HYPERTROPHIC OSTEOARTHROPATHY

The complete picture of hypertrophic osteo-arthropathy (HOA) involves clubbing and:
1. hypertrophy of the distal extremity and soft tissue proliferation
2. peripheral neurovascular disease; cyanosis, hyperhidrosis, and paresthesiae (variable)
3. bone or joint pain
4. proliferative periostitis
5. muscle weakness

More than 90% of true HOA occur with intrathoracic malignancies. Surgical vagotomy (but not atropine vagolysis) will reduce symptoms without changing the degree of clubbing (Lancet 1:343 '58).

About 30% of cases of HOA are associated with gynecomastia. In such cases, both manifestations are probably expressions of tumor endocrine activity.

IDIOPATHIC HYPERTROPHIC SUBAORTIC STENOSIS (IHSS)

A late or mid to late systolic murmur is characteristic (see also "MITRAL PROLAPSE"), but in many instances it will commence with the first heart sound. Ablation of the murmur with squatting from a standing position makes the diagnosis (the Nellen sign); with aortic stenosis squatting will intensify the murmur (Am H J 76:295 '68).

The intensity of the systolic murmur is characteristically increased after premature ventricular contractions and with isuprel or amyl nitrate administration (i.e. in those states in which the force of ventricular contraction is increased).

Some degree of mitral insufficiency is present in more than half of all cases, although it is usually of no clinical significance. The pathophysiology of this abnormality may involve asynchronous papillary muscle function consequent to asymmetrical ventricular hypertrophy.

The diagnosis of IHSS should be suspected when the typical murmur of aortic stenosis is maximal at the lower left sternal borner rather than in the aortic area and when it contains high pitched components at the left sternal border (indicating some concomitant mitral insufficiency).

IHSS is associated with the Wolff-Parkinson- White syndrome, aortic coarctation, and (rarely) small ventricular septal defect.

LEPROSY

Inability to straighten the little finger is an early sign of leprosy (Canizares, NYU). The claw hand deformity is a progression of this early change.

One of the earliest dermatologic signs is the formation of hypopigmented macules, suggesting vitiligo; these later become anhydrotic and anesthetic.

Superficial nerves become thickened, especially the posterior auricular, facial, ulnar, posterior popliteal, and sural nerves (similar thickening can be seen with interstitial hypertrophic neuritis and neurofibromatosis). Beading of corneal nerves (seen with the slit lamp) is a very early sign, especially of the lepromatous form. The X-ray finding of a calcified peripheral nerve is pathognomonic (Brit J Radiol 38: 796 '65). Tests of peripheral neuronal integrity are the production of a flare after intradermal injection of histamine (Pierini Test) or sweating after intradermal injection of pilocarpine. Bilateral VIIth nerve palsy is a common feature of leprosy.

In addition to St. Anne's sign (see ALOPECIA) there is usually alopecia of the beard. Puffy, enlarged ear lobes are very common with lepromatous leprosy. Facial infiltration results in the leonine facies.

Biologic false positive serologic reactions for syphilis occur with lepromatous leprosy, not the tuberculoid form.

Hansen's bacilli can be found in nasal septum scrapings late in the course of lepromatous leprosy, deep skin preparations are more likely to reveal organisms earlier in the course. When ear lobes are swollen, they are a likely site to reveal organisms.

Bacillary invasion of the testis is almost invariable. Urinary gonadotrophins are increased and urinary estrogens are decreased. In some 10 to 20% of cases, testicular atrophy and gynecomastia are noted clinically (Lancet 2:1320 '68).

Iris pearls are numerous, minute white spots in the iris adjacent to the pupil. They are best seen with the slit lamp and are pathognomonic of lepromatous leprosy. Episcleral nodules (encroaching on the cornea) where present are a source of Hansen's bacilli for laboratory confirmation of the diagnosis.

LESCH-NYHAN SYNDROME

The features of this disorder are X-linked hyperuricemia and neurological damage.

In the original cases neurological findings were:
1. mild mental retardation
2. choreoathetosis
3. bizarre behavior, particularly self mutilation

Uric acid overproduction has been demonstrated in patients and in cultured fibroblasts from affected individuals. The biochemical basis of this abnormality is an absence (or marked decrease) in phosphoribosyl transferase (PRT) activity. This enzyme is necessary for feedback inhibition of purine synthesis (Seegmiller, NIH).

PRT assays of the mothers of patients reveals a dimorphic cell population of full or no PRT activity, an expression of the Lyon hypothesis. PRT activity assay can be performed on cells obtained by aminocentesis at 18 weeks for counseling as to the continuation of pregnancy.

(CHRONIC MONOCYTIC) LEUKEMIA

Chronic monocytic leukemia occurs throughout adult life with peak presentation during the 6th decade. There is often an antecedent history of a chronic disease process such as tuberculosis. At the JHH, the incidence of this disorder is about 5 new cases per year. It is not often diagnosed early and should be suspected with:

1. refractory anemia
2. bleeding, especially with thrombocytosis or bizarre platelet morphology
3. unconfirmed clinical impression of a chronic disease
4. unexplained weight loss (as with other malignancies)

A long standing refractory anemia is a feature of the pre-leukemic phase. The smear may reveal red cell aniso-poikilocytosis and, at times, macrocytosis which suggests pernicious anemia or chronic hepatic disease. Large and/or bizarre platelets are characteristic.

Leukopenia with increasing monocytosis occurs during the leukemic phase. During this phase also:

1. 80% of patients have skin lesions - ecchymoses or petechiae (platelet abnormality) and blue tipped papules (specific, representing monocytic infiltration)

2. 50% have hepatosplenomegaly (though lymphadenopathy is rare)

3. 10 to 30 times the normal urinary lysozyme excretion (J Exp Med 124:921 '66)

Gingival hypertrophy occurs late in the leukemic phase or during the acute blastic phase, when the monocytes become increasingly immature and the disease terminates. Death may follow CVA, massive GI bleeding, or overwhelming infection.

(SYSTEMIC) LUPUS ERYTHEMATOSUS

Chronicity......episodicity......multiplicity

Bleeding, ranging from minimal purpura to massive hemorrhage occurs with SLE. The possible causes are:
1. acute arteritis
2. thrombocytopenia
3. thrombotic thrombocytopenic purpura
4. macroglobulinemia
5. uremia

6. hepatic disease
7. circulating anticoagulant
8. pancreatitis
9. drug therapy
10. septicemia

Suspect SLE with unexplained bleeding in women aged 20 to 40 years (Tumulty, JHH)!

In about 10% of cases the diagnosis is made after age 50. Although the incidences of fever, anemia, hyperglobulinemia, and pericarditis are similar to those in younger age groups, the following features are noted (MB Stevens, JHH):

1. the prodromal phase is longer and the severity less than in younger patients

2. Articular involvement is less frequent, occurring in about 50% of cases above age 50 versus 100% below age 30

3. Dermatologic changes are less frequent, occurring in about 33% versus about 80% below age 30

4. Pulmonary abnormalities, atelectasis and pneumonitis, are more frequent, occurring in about 20% of patients below age 30, 40% aged 30 to 40 years, and more than 50% above age 50

Sterile pyuria occurs with renal involvement. Bacituria must be confirmed by culture before antibiotic therapy is initiated, particularly in view of the frequency of abnormal drug reactions in these patients. On the other hand, patients with active SLE are prone to bacterial infections, one reason for which may be impaired phagocytosis of their granulocytes (Scand J Haem 6:348 '69).

Periungual erythema is a definite sign of activity. Subcutaneous periarticular nodules occur with active SLE as well as rheumatoid arthritis (MB Stevens, JHH).

Patients may complain of a persistent, aching sore throat which is intensified by swallowing. A posterior pharyngeal myositis occurs and may well prove to be characteristic of this particular disease.

The incidence of lymphoma is increased in patients with SLE.

Patients with SLE are often anergic to PPD, although they react normally to skin tests with extracts of trichophyton, candida, and other 'common' antigens (MB Stevens, JHH).

Lysosomes may mediate the tissue injury of immune reactions. Apropos, X-ray or ultraviolet radiation, endotoxin, and vitamin A labilize lysosomes, while cortisone, chloroquine, and chlorpromazine stabilize them. The skin lysosomes of patients with SLE are more 'fragile' than the norm. Intradermal application of vitamin A provokes a brisk inflammatory reaction which may eventually be refined into a sensitive screening or diagnostic test (Weissman, NYU).

MITRAL PROLAPSE

Mitral prolapse is a benign condition which is more common in women than in men. There is mitral insufficiency at the height of systole when the valve leaflets dome (or prolapse) into the left atrium. The murmur is, therefore, that of mitral insufficiency (i.e. systolic, extending through aortic closure), although it commences late in systole. It may be preceeded by one or more systolic clicks (Brit H J 28:488 '66).

Mitral prolapse occurs in almost all cases of Marfan's syndrome and is frequent in cases of Turner's syndrome. In these instances, the chordae are long and redundant.

In rare instances mitral prolapse may progress over the years to significant mitral insufficiency.

Late systolic murmurs are also seen with:

1. idiopathic hypertrophic subaortic stenosis

2. mammary artery bruits

3. post-infarction papillary muscle dysfunction (Burch syndrome), including some cases of sickle cell disease with localized, subclinical papillary region infarction

4. (rare) metastatic disease (choriocarcinoma in particular) involving the papillary muscle

(INFECTIOUS) MONONUCLEOSIS

The duration of the disease is 2 weeks or less in the large majority of patients. Clinical features are (Am J Hyg 71:342 '60):

1. fever, lymphadenopathy (100%)

2. pharyngitis (80%) (may be severe and exudative, atypical lymphs are abundant in the smear and are diagnostic when noted in working up a pharyngitis)

3. splenomegaly (40%)

4. periorbital edema (36%)

5. palatal enanthem (5%)

6. jaundice (5%) (fever is persistent which aids in differentiating this from infectious hepatitis)

Practically all cases have some degree of hepatic functional impairment by laboratory criteria. Unlike viral hepatitis, the alkaline phosphatase and LDH are elevated out of proportion to the transaminases. Fist percussion tenderness of the liver and cigarette distaste may be present as well.

Skin hypersensitivity reactions to ampicillin appear particular--ly common in patients with mono (Lancet 2:1176 '67).

The complications of mono are:
1. splenic rupture
2. obstructing laryngeal edema
3. pneumonitis (5%)
4. pericarditis, myocarditis (1%)
5. neurological complications (asceptic meningitis, Guillain-Barré syndrome) (1%), optic neuritis
6. hematologic complications (thrombocytopenia, hemolytic anemia) (1%)

MUCOPOLYSACCHARIDOSES

Types (McKusick classification):

I.   Hurler syndrome
II. Hunter syndrome
III. Sanfilipo syndrome
IV. Morquio-Ulrich syndrome
V.   Scheie syndrome
VI. Maroteaux-Lamy syndrome

The Hunter syndrome is X-linked, the remainder are autosomal recessive disorders. They are precisely differentiated by the type of mucopolysaccharide excreted in the urine, but they may be roughly distinguished on clinical grounds (McKusick):

Although dwarfism is a feature of the Hurler type, they are quite large as infants. Gargoylism is the 'classical' appearance. (An X-ray finding which may be specific is flattening of the mandibular condyle, A J Roent 86:473 '61).

The absence of corneal clouding distinguishes the Hunter type (although microscopic opacification can be found in some necropsy specimens of those surviving beyond the 3rd decade). Thickening of the skin at the level of the scapula along the posterior axillary line is a characteristic of the Hunter type.

Hirsutism is a feature of all types, although it is somewhat more marked in the Scheie type. Patients with this disorder have full physical and intellectual development but suffer from claw hand and foot, corneal clouding, and aortic insufficiency. Their faces are marked by bushy eyebrows.

The Morquio type is severely dwarfed. There is often odontoid hypoplasia, leading later to paraplegia or other cord problems.

The Sanfilipo type is without skeletal deformity but is characterized by severe progressive mental deterioration. The Maroteaux-Lamy type is generally of high intelligence and has long survival, although he is otherwise like the Hurler variety. A possibly pathognomonic feature of the M-L type is the presence of teeth eroding through the mandible on appropriate skull X-rays.

All of the mucopolysaccharidoses have peripheral Mittwoch cells, namely, lymphocytes with inclusion bodies, this, then, being the 'simplest' confirmatory lab test. These cells are most numerous in the M-L type.

In all of the mucopolysaccharidoses, there is a specific cellular deficiency of a protein involved in mucopolysaccharide degradation. Although types I, II, and III are different, the deficiency of I, V, and VI seems to be identical on the biochemical level. It is to be hoped that replacement therapy will be possible in the near future (Neufeld, NIH).

MULTIPLE MYELOMA

The 'classical' triad suggestive of myeloma is anemia, proteinuria, and bone pain. Clinical features are:

1. skeletal involvement
   present in about 90% of cases, pathological fractures are common; generalized osteoporosis may be an early finding; vertebral

body destruction leads to kyphosis and decreased stature, nerve root compression with vertebral body distortion leads to a variety of neurological symptoms and signs (indeed, next to cancer metastases, myeloma is said to be the commonest systemic cause of paraplegia (Waldenstrom, Multiple Myeloma, Grune and Stratton, New York, 1970).

2. immunologic abnormalities
production of abnormal (or otherwise deficient) antibodies promotes susceptibility to infections, patients may be overwhelmed by otherwise 'benign' bacterial or parasitic processes.

3. protein abnormalities
excessive globulin production may lead to the hyperviscosity syndrome with weakness, lethargy, venous plethora, mucous membrane bleeding, hemorrhagic diathesis, retinal vascular abnormalities, visual disturbances. Pyro- and cyro- globulins may be present.

4. anemia
invariably present; usually normochromic and normocytic, although macrocytosis may be present; an association between myeloma and pernicious anemia may occur (Acta Med Scand 172:195 '62). B_{12} studies in patients with dysproteinemias are indicated.

5. renal involvement
proteinuria is present in 80 to 90% of cases, with Bence-Jones proteinuria detected in about 50% of all cases; renal insufficiency is progressive; persistently alkaline urine with multiple myeloma (or other dysproteinemia) (in the absence of infection with urease splitters) suggests complicating renal tubular acidosis.

When hypercalcemia is a complication, ambulation is an important therapeutic adjunct; adequate hydration is essential.

Osteolytic lesions of myeloma are usually sharply marginated ("If the margin of the osteolytic lesion is fuzzy, the diagnosis of myeloma is unlikely" - Snapper, c.f. Waldenstrom, opp. cit.) Cranial vault masses of myeloma tend to "protrude, rarely, if ever, intrude" into the cranial cavity (Waldenstrom, opp. cit.).

A 20% coexistance of myeloma and primary carcinoma, most often of the GI tract, has been reported (CA 11:221 '58). Of 27 consecutive cases of myeloma in the JHH autopsy files (cases 20,000 to 30,000), 3 had coexistant GI cancers. Patients with a globulin spike should be thoroughly investigated for a coexistant cancer; it is impossible in many instances to unravel the dilemma of whether the protein abnormality preceded or was caused by the presence of a cancer.

MYASTHENIA

In myasthenia gravis the ocular and bulbar muscles are usually involved first, but the pupils are always normal. If ocular involvement is the sole manifestation of myasthenia for 18 months, systemic progression is most unlikely (JAMA 153:529 '53).

8 to 10% of patients with myasthenia have a thymoma (with which red cell aplasia or macrocytosis may be noted); almost all patients with thymoma have anti-muscle antibodies. Myasthenics with thymoma are much less responsive to prostigmine than those without thymoma.

Myasthenics must avoid curare, quinine, and quinidine (Walsh, JHH). Streptomycin, neomycin, kanamycin and the polymixins have curare-like effects in myasthenics.

Myasthenia may be associated with:
1. lupus
2. rheumatoid arthritis
3. hyperthyroidism

Myasthenic symptoms preceding the discovery of a malignancy is the Eaton-Lambert syndrome, which differs from classical myasthenia in the following ways:

1. It usually occurs in 5th decade males (as opposed to 3rd decade females)

2. there is proximal weakness, lower limb and pelvic musculature is most severely impaired, ocular and bulbar weakness is uncommon

3. peak strength develops (the defect is the poor release of AcCh to a single stimulus which can be overcome by repetitive stimulae). The N-M transmission pattern is diagnostic

The Eaton-Lambert syndrome is most often seen with oat cell carcinoma of the lung, although it has also been reported with GI, retroperitoneal, and other neoplasms. Other clinical features are dry mouth, hyporeflexia, and facial-upper limb paresthesias.

MYOCARDIAL INFARCTION (See also "ANGINA PECTORIS")

Recognition of an atrial infarction, with or without ventricular infarction, is important because of the danger of recurrent atrial arrythmia and the likelihood of atrial mural thrombus formation. The pertinent EKG findings are P-STa segment elevation with reciprocal depression or P-STa depression in association with an atrial arrythmia of sudden onset (The P-STa segment occurs after the P wave returns to the baseline and before the onset of the QRS complex, the P-Ta wave is generally buried within the QRS and is, therefore, only seen with A-V dissociation).

Infarction of the diaphragmatic surface of the heart may be associated with cutaneous hyperesthesia of the left supraclavicular region or with tenderness of the left trapezeal border.

Angina with syncope suggests posterior ischemia with decreased flow to the A-V node. With posterior infarction, intravenous atropine may decrease the risk of an arrhythmia.

Heart sounds may decrease in intensity during recovery from infarction, particularly with posterior infarctions (possibly because of reduced papillary muscle action) (Brit H J 30:835 '68).

When the clinical presentation suggests the possibility of an infarction, but the EKG is unremarkable except for a T vector change such that:
1. TI is less than TIII
2. TAVL is negative
3. TAV A & F is high

high chest leads are indicated, for in this situation there may be a small infarction of the high, lateral region (Brit H J 21:407 '59, 31: 623 '69).

Women with atherosclerotic cardiovascular disease have increased pregnancy loss (and compensatory increased number of pregnancies), independent of the occurrence of diabetes, during their child bearing years (Winklestein, Berkeley). In particular, there is an excellent correlation between "habitual abortion" and subsequent ASCVD.

If a patient suspected of having a myocardial infarction does not have an audible 4th heart sound, he does not have a myocardial infarction (Harvey, Georgetown)! The converse is not true. Apropos, a 4th heart sound of itself in a patient with an infarction does not imply cardiac failure.

Contrary to classical 'dogma' a gentle rectal examination is NOT contraindicated in acute myocardial infarction (see NEJM 281:238 '69). Indeed, the discovery of fecal impaction or the detection of occult bleeding is of extreme importance.

Post-myocardial infarction syndrome (Dressler's syndrome) presents as fever and pleuro-pericardial chest pain days to months after infarction. The features are:
1. pericarditis (with or without a small effusion)
2. pneumonitis
3. pleural effusion (usually unilateral, left)

The sed rate is elevated and there is mild leukocytosis. The chest film does not show congestion of the pulmonary vessels or Kerley B lines which aid in distinguishing this disorder from congestive failure and pulmonary infarction. Anticoagulants are contraindicated in this condition; steroids are therapeutic.

MYOTONIC DYSTROPHY

Blepharitis and decreased lacrimation occur early in the course of myotonic dystrophy, but the principle and characteristic ocular finding is multiple, minute anterior and posterior cortical and subcapsular lens opacifications, which are often polychromatic, being red, green, blue, and white. A posterior subcapsular cataract may evolve late in the course of the disease.

NEUROFIBROMATOSIS

The features of neurofibromatosis are:
1. cafe au lait spots (these should be larger than 1.5 cm in diameter and more numerous than 4)
2. cutaneous nodules, iris nodules
3. peripheral nerve tumors
4. adrenal tumors (pheochromocytoma)
5. orbital and intracranial tumors (meningioma, acoustic neuroma, optic nerve glioma)

In children the following may be found:
1. inequality of the length and bulk of limbs
2. localized gigantism of fingers
3. congenital glaucoma
4. subperiosteal defects in long bones, rib anomalies
5. vertebral column defects
6. pseudarthroses

Axillary freckles may be pathognomonic of neurofibromatosis (Ann Int Med 61:1142 '64).

OSLER-QUINCKE DISEASE (Hereditary angioneurotic Edema)

This disorder presents as PERIODIC ACUTE PERIPHERAL EDEMA and ABDOMINAL PAIN.

The underlying biochemical abnormality is the absence (or decreased activity) of a specific $alpha_2$ globulin which inhibits C'l esterase and which also inhibits the kallikrein-kinin activating system (NEJM 272:649 '65). (C'l esterase normally inactivates some complement components.)

Although the protein defect is present throughout life, symptoms rarely appear before age 6 and are often delayed until adulthood.

Edema is the basic pathophysiologic disturbance, involving the skin, upper respiratory tract, and gastrointestinal mucosa. Skin edema tends to be circumscribed and without pitting. Laryngeal edema is a frequent terminal event. GI mucosal edema leads to acute abdominal pain which may mimic virtually any GI or GU disorder in its clinical presentation; porphyria must, in particular, be ruled out in the differential diagnosis. Any level of the gastrointestinal tract may be locally involved. Skin edema generally precedes or accompanies edema at the other sites.

Attacks are often cyclical and of great regularity. The provoking stimulus is almost always unknown, although in some instances trauma can be implicated. Women are generally free of attacks during pregnancy.

Antihistamines relieve skin edema, corticosteroids (and at times testosterone) ameliorate abdominal pain.

OSLER-WEBER-RENDU DISEASE

Osler-Weber-Rendu disease may be clinically mimicked by the CRST syndrome. The incidences of cirrhosis and peptic ulcer disease are increased with the OWR disease.

Telangiectases in this condition vary with time, increasing with age and during pregnancy. Bleeding is uncommon during childhood. Pulmonary A-V fistula may occur, the classical physical sign being a bruit which intensifies during inspiration. Even the roentgenographically smallest pulmonary A-V fistula will likely have serious clinical implications, leading to cyanosis and marked increase in the cardiac output.

Globular retinal venous varices are indicative of hereditary hemorrhagic telangiectasia. Rupture of conjunctival telangiectases causes 'bloody' tears.

PACHYDERMOPERIOSTOSIS (Touraine-Solente-Gole Syndrome,
   Arch Derm 94:594 '66)

Pachydermoperiostosis is a familial disorder with decreased penetrance in women which usually appears in the 3rd or 4th decade. The clinical features are:
   1. clubbing
   2. palmar keratoses
   3. generalized bony periosteal elevation
   4. an acromegaly-like picture of hyperhidrosis, excessive sebaceous gland activity, macroglossia, and enlargement of the hands.

Palmar keratoses are short, interrupted, and linear in arrangement. Redundant longitudinal scalp folds are characteristic.

The appearance of this disorder beyond the 4th decade suggests lung carcinoma as a cause. Pachydermoperiostosis, when secondary to a neoplasm, is probably of the clinical spectrum, including hypertrophic osteoarthropathy, which represents the effects of tumor endocrine activity.

PAGET'S DISEASE OF BONE

Paget's disease of bone may occur in as much as 3 or 4% of individuals over age 40. In addition to the danger of sarcomatous change, it is well to recall that involved bony zones are regions of predilection for implantation of metastatic carcinoma.

Although there is bony thickening and intensive osteoblastic activity, involved bones are structurally weak because of disruption of the regular supporting matrix. Pathological fractures are common. Osteoblastic activity is reflected by high serum alkaline phosphatase activity and high urine hydroxyproline levels.

Involved bony regions are quite vascular. With extensive bony involvement there is an A-V fistula effect with widened pulse pressure. At times, patients may present with congestive heart failure (though in this situation the circ time will be short). Examination of the eye may reveal angioid streaks, papilledema, or optic atrophy; senescent macular degeneration is common.

Neurological signs are secondary to entrapment of cranial nerves exiting from an enlarging skull. The auditory portion of the VIIIth nerve is most often involved. Cochlear distortion may contribute to deafness.

As with diabetes, arteriosclerosis occurs in patients with this disease earlier and more extensively than in nonaffected persons.

Iliopectineal line thickening (the brim sign) is an early X-ray finding (Am J Roent 90:1267 '63). Osteoporosis confined to the parietal bones may be the earliest skull X-ray finding.

PANCREATIC DISEASE

Carcinoma:

Cytological examination of duodenal aspirate after pancreatic stimulation is a useful diagnostic technique.

With local invasion a bruit may be audible over the splenic artery.

The 'classical' stool of invasive pancreatic neoplasm is the greasy, silver stool (the pale stool of obstruction admixed with a small amount of blood).

Ptyalism is sometimes seen with pancreatic carcinoma (Osler).

Pancreatitis:

Pancreatitis is associated with biliary tract disease, alcoholism, and:

1. post-partum state
2. chronic adrenal steroid therapy
3. hyperparathyroidism
4. essential hyperlipemia

The fever of pancreatitis is generally low grade unless there is abscess formation or extensive pancreatic necrosis. In this latter condition, paracentesis may produce "beef-broth" or "prune-juice" ascitic fluid (an admixture of blood, cell detritus, and pancreatic secretions).

The pain pattern of acute pancreatitis is quite variable, and frequently it is marked by severity. Patients often flex the trunk maximally for relief; one such flexed posture, the 'Mohammedan prayer position' is a diagnostic clue (see Press Med 67:1543 '59).

Physical signs may be few, despite the intensity of the abdominal pain. It is said that the pancreas can be palpated by placing the patient in the right decubitus position and attempting to insinuate the hand between the medially displaced stomach and the laterally placed spleen. Tenderness in this region would appear to indicate pancreatitis (much as tenderness over McBurney's point may help in the diagnosis of appendicitis).

An elevated left hemidiaphragm with linear basal streaking or blatant atelectasis is present in about 30% of cases of acute pancreatitis. In about 10% of cases, pleural effusion occurs, being generally restricted to the left side. Pleural fluid amylase may be quite high in this situation.

Pancreatic disease should be kept in mind when unexplained erythematous, painful, subcutaneous nodules are present (Knopf, NYU).

PARATHYROID DISEASES (See also "CALCIUM")

The parathyroid is unlikely to be the cause of hypercalcemia when both the venous CO_2 and urine calcium are elevated. A fourth of patients with hyperparathyroidism have normal urine calcium; in the remainder, the increase is less marked than with other causes of hypercalcemia (particularly the milk aklali syndrome). Several instances of normocalcemic hyperparathyroidism have been reported (Am J Med 47:384 '69).

Although less than 5% of patients with renal calculi will be found to have primary hyperparathyroidism, every individual with renal calculi (and especially those with recurrent calculi) should be thoroughly worked-up for this treatable condition.

McCallum (Bull JHH 16:87 '05):

At an autopsy in the case of a young man age 26, who had suffered for several years from symptoms of a chronic nephritis and on whom 2 years ago the operation of decapsulation of the kidney had been performed in the Johns Hopkins hospital, there was found in the right side just below the lower pole of the thyroid and quite separate from it a round, smooth mass enclosed in a delicate capsule and lying loose in the connective tissue.

The anemia of hyperparathyroidism only occurs with marked bony changes. When bone films are normal, other causes for a coincident anemia must be sought. The radial and palmar surfaces of the middle phalanges of the dominant hand are the earliest (and most consistently) regions of bony involvement. Resorption in phalanges is generally subperiosteal, in the long bones subendochondral.

Pseudo- and pseudo-pseudo- hypoparathyroidism are probably the same disorder, both are characterized by:
1. short stature
2. increased facial skin folds - "prune" facies
3. shortening of the lateral metacarpals
4. calcification and ossification in fascial planes, ectopic calcification, cataracts
5. lack of response to parathormone (see also "BASAL CELL NEVUS SYNDROME")

In addition, many patients with pseudohypoparathyroidism have anatomic and physiologic abnormalities of the palate, such that the palate is elongated and flat, there is decreased taste sensation for bitter and sour taste objects (taste buds for this modality are abundant in the palate, Ann Int Med 71:791 '69).

Hypoparathyroidism may present in children (and rarely in adults) as steatorrhea, the pathophysiology of which is as yet uncertain.

Hypoparathyroidism may also present in children as lamellar or zonular cataracts (rather than subcapsular cataracts). At times these opacifications regress with correction of the metabolic defect.

Patients with hypoparathyroidism very often have sparse eyebrows.

About 50% of patients with hypoparathyroidism who develop severe keratoconjunctivitis will develop or have concomitant Addison's disease (A J Ophth 54:660 '62).

PEPTIC ULCER DISEASE (See also "GASTRIC ULCER - GASTRIC CARCINOMA")

A somewhat fanciful sounding, but interesting, differential sign of duodenal ulcer disease is the Warthin frown sign, which was occasioned by the observation that 3 or more longitudinal forehead furrows are present in 90% of patients with peptic ulcer disease (4 or more in 65%) in contrast to 2 or fewer furrows in 95% of non-ulcer controls (JAMA 205:470 '68).

The incidence of duodenal ulcer is increased with:
1. hepatic cirrhosis
2. chronic pancreatitis
3. cystic fibrosis
4. rheumatoid arthritis
5. multiple endocrine adenoma syndrome
6. hyperparathyroidism
7. (?) chronic obstructive lung disease
8. carcinoid syndrome
9. Osler-Weber-Rendu disease

In contrast, duodenal ulcer rarely occurs with pernicious anemia or with ulcerative colitis (prior to steroid therapy).

Although physical exam is generally unrevealing, with active ulcer disease there may be:
1. localized tenderness to percussion of the epigastrium (Mendel's sign)
2. localized epigastric hypothermia

Spices do NOT increase acid secretion. Dietary restrictions should be individualized to food which cause the individual patient dyspepsia, i.e. pizza need not necessarily be forbidden. Frequency of feeding is the most important aspect of dietary management.

Post-bulbar duodenal mucosa is abnormal in appearance whenever there is a duodenal ulcer, independent of ulcer location within the bulb. When there is some question about the presence of a duodenal ulcer, and the post-bulbar mucosa is normal on spot films it is safe to state that no ulcer is present (Dreyfuss, MGH).

PERICARDIAL DISEASE

In addition to infectious and granulomatous diseases, pericarditis can be associated with:
1. lupus erythematosis, scleroderma
2. uremia
3. lymphoma, leukemia
4. malignant melanoma
5. lung cancer, breast cancer
6. active or latent gout (Lancet 1:21 '63)

TB pericarditis often presents as an atrial arrhythmia.

The pericardial friction rub is classically 3-phased, occurring with atrial systole, ventricular systole, and ventricular relaxation. A to-and-fro scratchy sound in the pulmonic area occurs in some instances of hyperthyroidism (Means-Lerman scratch, Am H J 8:55 '32) and may at first be mistaken for a pericardial rub.

Simultaneous atrial and ventricular electrical alternans is uniquely related to pericardial effusion and indicates long standing effusion with development of tamponade. This sign is particularly common with large, malignant effusions. The electrocardiograph changes are secondary to excessive cardiac motion, which can be demonstrated with facility with ultrasound techniques (Circ 34:611 '66).

Development of a pericardial rub in a patient with a transvenous pacemaker indicates myocardial perforation. Chest pain and signs of diaphragm or abdominal muscle electrical stimulation are usually present. Surprisingly, this complication usually does not present as an emergent catastrophe!

PHEOCHROMOCYTOMA (Ann Int Med 65:1302 '66)

Half of cases have persistent hypertension, half have paroxysmal hypertension. In the former group, women predominate 2:1, in the latter M = F. About 10% of pheochromocytomas are malignant.

Pheochromocytoma should be suspected with:
1. hypertension and diabetes
2. headache, fever, sweats, nausea
3. hyperthyroid like picture with normal T4
4. hypertension or shock with surgery
5. acute anxiety attacks
6. acute heart failure without previous cardiac disease
(there may be an associated myocarditis, NEJM 274:1102 '66)
7. severe pre-eclampsia in multips
8. "unexplained" arrhythmias
9. hypertensive retinopathy with normal vessels

Pheo's are associated with:

    1. amyloid secreting medullary carcinoma of thyroid (Ann Int Med 63:1027 '65)

    2. Cushing's disease (some pheo's produce ACTH)

    3. multiple endocrine adenoma syndrome

    4. polycythemia (secondary to erythropoetin secretion)

    5. neurofibromatosis

    6. cholelithiasis (in about 30% of the paroxysmal hypertension type, 10% of the persistent type   Mayo Clin Proc 39:281 '64)

The glucagon provacative test is helpful in making the diagnosis (Ann Int Med 66:1091 '67).

In some uncommon instances, symptoms may be precipitated by postural changes which cause tumor compression such as lifting heavy objects or stooping forward to tie a shoe lace.

PNEUMONIA

Pneumonia may present in elderly patients as congestive heart failure.

Delerium and confusion may complicate the course of alcoholics with pneumococcal pneumonia - BEWARE THE BEADED BROW in hospitalized alcoholics.

Inappropriate ADH secretion can occur with pneumonia alone.

Protoplast pneumonia may be associated with pneumococcal pneumonia and should be suspected when the chest film is disproportionately bad. Bullous myringitis with or without pharyngitis implicates the Eaton agent. (Eaton agent pneumonia is particularly common among military recruits).

Common associations:

    1. alcoholism - klebsiella (sputum is extremely tenacious, the chest film classically reveals upper lobe involvement with bulging fissures).

    2. aspiration - pseudomonas

    3. lung cancer, debilitation, deep intravenous lines - staph (tachypnea without pleurisy is present, staph pneumonias also seem to occur more commonly than anticipated in pregnant women)

Possibilities accounting for delayed resolution of a bacterial pneumonia are:

    1. improper antibiosis

    2. underlying carcinoma

    3. immunologic abnormality (common in alcoholics)

    4. drug reaction

    5. development of systemic complications, viz: endocarditis, abscess, empyema, pericarditis, arthritis, etc.

When the chest film indicates pneumonia and the physical reveals mouth ulcers, consider histoplasmosis.

Pneumonia due to aspiration may involve the anterior and posterior segments of the right upper lobe when the patient has been in the right decubitus position during the event, a not infrequent occurrence in alcoholics sleeping off stupors.

POLYCYTHEMIA

Causes of abdominal pain in patients with polycythemia may be:
1. splenic infarction
2. stretching of the hepatic capsule
3. peptic ulceration (mucosal vessel thrombosis may be involved)
4. mesenteric thrombosis

A swollen, red ocular caruncle is characteristic, though of itself not diagnostic of polycythemia. The lids often have a dusky, cyanotic coloration.

The chest X-ray in polycythemia reveals pulmonary vascular distention which is visible out to the periphery in as much as 60 per cent of cases. Many vessels in the outer fields are seen 'end on', presenting a mottled appearance. Kerley B lines are not seen, which helps to distinguish these changes from those of congestive heart failure. Plate-like atelectasis or more blatant indications of pulmonary infarction may be noted (Clin Radiol 12:276 '61).

PORPHYRIA - ERYTHROPOETIC TYPES

Congenital erythropoetic porphyria is a rare disorder with recessive inheritance. The urine is port-wine color. There is hirsuitism and facial hypertrichosis as well as marked sun sensitivity. With exposure to sunlight, bullae appear; with time, scarring and disfiguration ensue. This disorder should be suspected when an infant cries in sunlight.

Erythropoetic protoporphyria is a dominant trait which appears commonly in the general population in a partial form. Although there is an acute burning reaction to sunlight, chronic skin changes, scarring, and disfiguration do not occur. Life span does not appear to be reduced in these individuals. The diagnosis is suggested by sun sensitivity but is confirmed by erythrocyte protoporphyrin studies (the urine is not colored). The mechanism of this disorder may, perhaps, involve photooxidation of skin lysosome membranes, in which case therapy might be directed towards stabilizing this membrane (see systemic LUPUS ERYTHEMATOSIS).

PORPHYRIA - HEPATIC TYPES (Tsudy, NIH)

Nervous sytem dysfunction involves:

1. autonomic neuritis - causing abdominal pain, postural hypotension, tachycardia, et cetera

2. peripheral neuritis - sensory abnormalities usually precede motor abnormalities and may occur exclusive of them

3. cranial nerve palsies

4. neuroses or acute psychoses - depression is the commonest mental aberration

Acute attacks may be precipitated by excessive ethanol intake and by a variety of drugs such as barbiturates, sulfonamides, griseofulvin, and less often meprobamate, glutethimide, mesantoin, and orinase. Demerol, chlorpromazine, antibiotics, and aspirin can be used safely.

Acute attacks are also precipitated by starvation, acute infectious processes, and variations in sex steroids, particularly estrogen (and including those induced by oral contraceptive agents). Women are most liable to acute attacks during the premenstrual phase when endogenous estrogen levels are highest.

During an acute attack BSP retention is increased. In about 40% of cases cholesterol levels are slightly elevated. Thyroid binding globulin is often increased, falsely elevating PBI and T4 levels, which may be particularly confusing when an acute attack mimics hyperthyroidism.

Excessive ADH secretion may accompany an acute attack leading to water intoxication with hyponatremia and, at times, tetany secondary to increased magnesium excretion. Hypothalmic disorganization has been demonstrated in autopsy studies of some cases with this complication.

The metabolic defect is increased delta AMINO LEVULINIC ACID SYNTHETASE synthesis in the liver. Overproduction of this enzyme is prevented by glucose. Therapy, therefore, should include intravenous glucose administration with insulin coverage followed later by a high carbohydrate diet. Chlorpromazine is valuable during an acute attack because it reduces anxiety by its tranquilizer effect and relieves abdominal pain by its ganglionic blocking activity.

PSEUDOXANTHOMA ELASTICUM

Pseudoxanthoma elasticum is inherited as an autosomal recessive. Features involve:

1. the skin
the typical facies is one with prominent nasolabial folds and excessive wrinkling about the chin
virtually all patients have some degree of lax skin about the neck or axilla
the characteristic changes (which may begin in youth) are those of redundant, wrinkled, "moroccan leather" skin in which there are multiple, irregular yellow plaques (yellow plaques are also found along the inner lip, buccal membranes, hard palate or in the gastric mucosa)

2. the cardiovascular system
vascular calcification is characteristic, arterial obstruction may be present at many levels of the arterial tree; about half of patients have hypertension; intermittent claudication or coronary artery insufficiency may, likewise, occur.
infrequently, patients have aortic insufficiency secondary to enormous dilatation of the aortic root

3. the ocular fundus
physical examination will reveal angioid streaks of the fundus in as much as 85% of cases
easily visualized choroidal vessels are noted in heterozygotes
chorioretinal scarring and hemorrhagic macular degeneration may seriously decrease visual acuity (indeed, skin biopsy is indicated in any young individual with hemorrhagic macular degeneration)

There is often an history of upper GI bleeding, which may be due to mucosal arteriolar degeneration. (Platelet size is increased in peripheral smears and may, possibly, indicate some thrombocytopathia contributing to the hemorrhagic diathesis). Non-specific abdominal complaints are common, and may suggest peptic ulcer. The GI series is usually negative, despite the fact that bleeding may be protracted.

In addition to vascular calcification, X-rays may reveal soft tissue calcium deposition.

Patients often initially seek plastic surgery consultation for cosmetic reconstruction of their redundant, lax facial and neck skin.

PSORIASIS (Psoriatic Arthropathy)

Psoriatic arthropathy occurs in slightly less than 10% of cases of psoriasis. Although it generally follows long after the onset of cutaneous changes, it may PRECEDE skin changes as much as 20% of the time.

Psoriatic arthropathy differs from rheumatoid arthritis in that constitutional signs are minimal, morning stiffness is rare, rheumatoid factor is not present, and subcutaneous nodules are not found. Other clinical features are:

1. frequent involvement of the DIP joints of the hands
2. asymmetrical, spotty joint involvement
3. involvement of more than one joint per finger (leading to the so-called "sausage finger" deformity)
4. invariable, concomitant nail pitting
5. hyperuricemia (about 40% of cases)

The X-ray features of psoriatic arthropathy are:
1. tuft resorption
2. discrete erosions at joint margins
3. spotty osteoporosis
4. bony ankylosis
5. paraspinal ligamentous ossification (syndesmephyte formation)

In hand films of good quality, finger nail integrity can be assessed. In a puzzling case, normal nails rule against psoriatic arthropathy.

(CHRONIC) PULMONARY DISEASE

Atrial flutter is a relatively frequent arrhythmia in cor pulmonale secondary to chronic obstructive pulmonary disease. Atrial fibrillation, however, is much less common in this setting, and its occurrence should suggest concomitant left-sided heart disease.

Hospitalized chronic bronchitics with acute decompensation often die suddenly as their blood gas determinations indicate improvement. Intensive monitoring and nursing care must not be relaxed during this phase despite gratifying clinical improvement.

The electrocardiographic findings of pulmonary heart disease represent cardiac position changes, the effect of adjacent hyperinflated lung, and crista superventricularis hypertrophy (Am J Cardiol 11:622 '63). These are:

1. small R waves in all leads
2. S1, S2 or S1 S2 S3 and persistence of S waves across the precordium

Individuals with chronic bronchitis have a high incidence of recurrent respiratory tract infections (particularly bronchitis) during the school years. This correlation is found both retrospectively and prospectively, that is, the innocent bronchitis of childhood may herald significant later disability. In some individuals with severe chronic bronchitis at an early age, serum alpha$_1$ antitrypsin activity is deficient (Am J Med 45:220 '68). Since this enzyme is the main component of the alpha$_1$ globulins, deficiency of this protein can be inferred from routine serum protein electrophoresis by a diminution of the alpha$_1$ spike. This protein deficiency is inherited (naturally!), although the specific genetic pattern is not entirely worked out.

E J M Campbell (Thorax 24:1 '69) lists the following physical signs of disturbed chest wall mechanics in patients with chronic bronchitis, which, in addition to the 'typical' physical findings, may aid in 'quantitating' the clinical status of these individuals:

1. reduction of the length of trachea palpable above the sternal notch

2. tracheal descent with inspiration

3. use of accessory muscles of respiration (sternomastoids, scalenes)

4. excavation of the supraclavicular and suprasternal fossae with inspiration

5. jugular venous filling during expiration

6. loss of the "bucket handle" movement of the mid points of the upper ribs (exaggeration of the upward, outward motion of the anterior ribs)

7. paradoxical movement of the costal margins (Hoover's sign)

Perhaps the very earliest chest X-ray finding in chronic bronchitis is blurring and indistinctness of the vascular markings in comparison with previous films. This change reflects thickening of adjacent bronchial walls with long standing inflammation. Though subtle, when this change is noted, significant pulmonary dysfunction is already present.

PULMONARY EMBOLUS

The great majority of the cases occurring in the hospitalized patient have subtle presentations: (Tumulty, JHH)

1. transient restlessness or anxiety, which may be misinterpreted as "nerves"

2. transient elevations of pulse, respiratory rate, or temperature, alterations of blood pressure

3. sudden weakness, fainting, or dizziness, which may initially suggest malingering

4. sudden development of paroxysmal arrhythmia

5. transient wheezing respirations

6. transient angina pectoris

7. irritating cough

8. sudden onset of congestive heart failure

9. upper abdominal pain; unilateral or bilateral shoulder pain

10. unexplained leukocytosis or increased bilirubin

11. basal "pneumonia, " especially with a streaky X-ray appearance

Roentgen signs are said to be present on plain chest films in more than 50% of cases in the first 24 hours (Clin Radiol 18:301 '67, Sem Roent 2:389 '67). When infarction has not occurred, the following may be seen:

1. unilateral increase in radiolucency
2. restriction of ventilation on the involved side - elevated diaphragm, discoid atelectasis
3. dilatation of the central pulmonary arteries
4. abrupt amputation of a proximally dilated pulmonary vessel

Infarction tends to occur in regions of lung with reduced collateral supply, lower lobes are involved more often than upper lobes; in all cases the infarction lies along a pleural surface and frequently it is at the junction of two pleural surfaces (apropos, fissures are pleural surfaces!). The proximal margin of the infiltrate on the chest film is rounded (Hampton's hump) and has a slightly diffuse margin. A small pleural effusion is present on the involved side in about half of all cases; as infarction typically involves the cortophrenic sulcus region differentiation is difficult (free fluid at the sulcus is generally concave to the hilus in contrast to the infarction hump which is convex).

With the appearance of an infiltrate on the chest film, THE FINDING OF WIDENING OF A PREVIOUSLY NORMAL PULMONARY ARTERY SEGMENT WITHIN 24 HOURS OF THE CLINICAL PRESENTATION OF RESPIRATORY DISTRESS DIFFERENTIATES PULMONARY EMBOLIZATION FROM PNEUMONITIS! (Identical X-ray technique and positioning is mandatory for this type of serial comparison).

Isoproterenol by intravenous administration may be useful in the emergent treatment of acute pulmonary embolus to decrease pulmonary vascular resistance and increase pulmonary blood flow.

PULMONARY HYPERTENSION

When physical signs and radiographic features of pulmonary hypertension are associated with an history of syncope with exercise, there is excellent reason to believe that there is pulmonary vascular obstructive disease rather than a left to right shunt (Perloff, Georgetown). In this situation, exercise precipitates acute right-sided failure since the increased venous return cannot be transmitted through the pulmonary bed. As a result there is an acute decrease of systemic output and syncope occurs. A sudden decrease of systemic output does not occur with a shunt because of a reversal in flow direction with right side decompensation. With mitral stenosis, there is generally fatigue precluding further exercise before the stage of syncope is reached.

Pulmonary hypertension leads to narrow splitting of P2. Wide splitting indicates right heart decompensation.

The cause of pulmonary hypertension is not always clarified by auscultation of the heart, because the murmurs of a patent ductus or of a ventricular septal defect may not be present. However the triad of a closely split P2, a pulmonic ejection sound, and a left-sided gallop indicate that a "silent" left to right shunt is causative (Harvey, Georgetown).

When a patient with advanced pulmonary hypertension presents with the sudden onset of severe chest pain radiating to the back, pulmonary artery dissection, though rare, should be considered.

REITER'S SYNDROME

Reiter's syndrome is a triad of:
ARTHRITIS
CONJUNCTIVITIS and
URETHRITIS-PROSTATITIS

It occurs some 20 times more often in men than in women and follows sexual exposure (in some rare cases it may follow acute diarrhea as in the WWII Finland dysentery epidemic).

Arthritis is recurrent and involves the periphery or the spine.

1. Asymmetrical joint involvement is characteristic; most often clinical arthritis is limited to the large joints of the lower extremity, but the small joints of the hands and feet, the sacroiliac joints, and the vertebral bodies may be involved as well. (Fusion of the sacroiliac joints does not, however, occur).

2. The synovial fluid is inflammatory, but glucose levels are normal or only minimally decreased. Complement is high. Smears may reveal large macrophages with ingested leukocyte fragments.

3. X-rays reveal periostitis and periarticular erosions. Periostitis of the os calcis (with heel pain) is a characteristic feature.

65% of patients with Reiter's syndrome have skin lesions:
1. (most often) hyperkeratosis of the palms and/or soles (keratoderma blenorrhagica)
2. circinate balanitis
3. oral mucosal ulcers, stomatitis
4. separation and heaping up of nails (similar to that of psoriasis but without prior geographic pitting)

Ocular findings are:
1. conjunctivitis (possibly mucopurulent)
2. anterior uveitis

Up to 30% of males with acute anterior uveitis are found to have or develop Reiter's syndrome (Perkins, Uveitis and Toxoplasmosis, London, 1961).

Reiter's syndrome is associated with aortic insufficiency and with first or higher degrees of heart block.

A quarter of patients with Reiter's syndrome have elevated psittacosis titers, indeed the etiology of this disorder appears to involve a Bedsonia organism.

RENAL DISEASE

Pregnancy is an excellent test of renal function. A pregnancy without toxemic changes provides at least some historical evidence of normal renal function at that phase of life.

The normal temperature response to infection may be absent in uremia; hypothermia is a feature of the terminal stage.

Renal vein thrombosis is often associated with underlying renal or multisystem disease and should be considered when such patients have rapid deterioration of renal function with oliguria. Renal vein thrombosis is frequently complicated by pulmonary thromboembolism, conversely, when pulmonary embolism is associated with proteinuria and microscopic hematuria, renal vein thrombosis is to be ruled out.

In cases of perplexing hepatic disease it is well to recall that hypernephroma (without metastatic spread) may present with slight jaundice (chiefly indirect bilirubin elevation), BSP retention, and other parameters of hepatic dysfunction (hypoprothrombinemia; decreased albumin; increased alpha$_2$ globulin, alkaline phosphatase). Splenomegaly may also be present. The liver biopsy reveals a nonspecific reactive hepatitis. With resection of the primary, these laboratory changes revert to normal; however, they again deteriorate if metastases develop (Proc Mayo Clin 45:161 '70).

When chronic renal disease is associated with peptic ulcer, analgesic (abuse) nephropathy should be considered. Calcification of the renal papillae may at times be noted in the plain abdominal films of patients with this type of renal disease (Clin Radiol 10:394 '68).

In pure renal disease, the BUN:creatinine ratio should be at least 10:1. When the ratio is less, some pre-renal process is present (such as hepatic disease, blood in the GI tract, deranged protein metabolism, etc.).

A hot bath will often relieve the pain of acute renal colic and should be employed whenever narcotic analgesics are either unavailable or contraindicated.

RHEUMATIC HEART DISEASE

Head bobbing occurs with aortic insufficiency, but head swaying (lateral movements) indicates severe tricuspid insufficiency.

Jugular venous distention in tricuspid insufficiency is asymmetrical, being more prominent on the right. The A wave increases with inspiration (whereas arterial pulses decrease slightly). Most men are not aware of any jugular venous distention they may have, but if they are specifically questioned about the appearance of the neck during morning shaving, an enlightening history may be obtained (Perloff, Georgetown).

Red cell fragmentation can occur with severe rheumatic valvular involvement as well as with prosthetic intracardiac materials. Probably related to this is the increased incidence of cholelithiasis in patients with mitral valve disease.

Mitral stenosis is the leading presumptive diagnosis when pulmonary hypertension is detected in an adult.

Lung perfusion in patients with mitral stenosis can be followed quantitatively with the lung scan.

Presystolic accentuation of the murmur of mitral stenosis implies that the anterior leaflet is mobile and that, therefore, a simple commissurotomy can be performed (Criley, JHH).

Hoarseness is a complication of mitral stenosis and is due to paralysis of the recurrent laryngeal nerve which is compressed between the aorta and an enlarged pulmonary artery (A H J 50:153 '55).

Paradoxical expansion of the left atrium in a patient with mitral valve disease (which is appreciated by chest fluoroscopy) indicates at least some degree of mitral regurgitation. Conversely, when ventricular pulsations are forceful, absence of paradoxical expansion is practically certain evidence of pure mitral stenosis (Radiol 54:1072 '65). Presumably, this information can also be obtained by ultrasound techniques which can also be utilized for 'direct' inspection of mitral valve motion.

RHEUMATOID ARTHRITIS

Rheumatoid arthritis can appear in concert with polyarteritis, systemic sclerosis, or lupus erythematosus. It is, however, NOT related to ankylosing spondylitis, Reiter's syndrome, psoriatic arthropathy, or colitic arthropathy.

The joint involvement of rheumatoid arthritis is:
Symmetrical,
Additive, and
Deforming

An early sign of rheumatoid arthritis is pain on compression of the head of the 5th metacarpal. With progression of the disease, patients prefer shoes to bare feet (Gessner, NYU). Wrist tenderness is an equally reliable but somewhat later noted sign.

Constitutional features, weight loss, low grade fever, and anemia indicate the systemic nature of the disease. Extra-articular features are:

1. subcutaneous nodules - 25%
2. lymphadenopathy - 25%
3. palmar erythema, without concomitant hepatic disease -
10%
4. vasculitis - 10%
5. splenomegaly - 5%
6. episcleritis, scleritis, sclero-keratitis (rarely scleromalacia perforans and very rarely uveitis)
7. keratoconjunctivitis sicca 10% - 15%

Leg ulcers occur in about 10% of patients and may be the result of a localized vasculitis.

Amyloidosis is a complication which should be suspected when patients develop renal disease, cardiac failure, or GI bleeding.

Histologic evidence of pericarditis is present in about 40% of patients at autopsy. Recurrent pleuritis is a clinical feature of rheumatoid arthritis. When nodular pleuro-pulmonary disease occurs, there may be pleural effusion. Very low glucose levels in the pleural fluid, though not specific, support the diagnosis; RA cells and high LDH may also be found (Dis Chest 53:202 '68).

Nodular changes on chest film are suggested as rheumatoid when the distal clavicular tip is resorbed. This clavicular change also occurs with scleroderma and hyperparathyroidism.

Platelets are often increased in number when the disease is active. About 75% of patients have a + RA factor, less than 25% have + LE preps.

Chronic bronchitis in men and bronchiectasis in men and women occur far more often in patients with rheumatoid arthritis than those with degenerative joint disease, usually antedating the onset of the joint symptoms. Bronchiectasis was noted in about 3% of some 500 patients with rheumatoid disease as compared with .3% of 300 with degenerative changes (Q J Med 36:239 '67).

A significant number of cases of the idiopathic uveitis syndrome of childhood (band keratopathy, iridocyclitis, secondary cataract) later develop Still's disease.

SCHIZOPHRENIA

Prognostic features (Stevens, Phipps Clin, JHH):
1. clear precipitating factor
2. acute onset
3. concomitant depression
4. marriage
5. good premorbid adjustment
6. no emotional blunting
7. expressed guilt
8. confused on admission

When 6 or more of these factors are present, recovery is likely, but when 5 or fewer are present the prognosis is poor.

Suspect schizophrenia in adolescents when good school performance suddenly and unexpectedly deteriorates (Frank, JHH).

The acute onset of visual hallucinations in a hospitalized schizophrenic indicates drug toxicity.

SCLERODERMA

F:M = 4:1. Peak age of presentation 20 to 40 years. Since the initial symptoms are often quite vague, an early impression is often hypochrondriasis!

Scleroderma should be suspected with (Tumulty, JHH):
1. uremia + malignant hypertension
2. unexplained dyspnea, pulmonary hypertension
3. pulmonary fibrosis in a young patient
4. telangiectasia (buccal mucosa is often involved), brawny pigmentation of the face, vitiligo of the chest
5. pericarditis, 'idiopathic' myocarditis
6. dysphagia, dyspepsia, hiatus hernia, duodenal ulcer, entero-colitis

Pulmonary connective tissue proliferation leads to alveolar destruction, cyst formation, emphysema, atelectasis, and bronchiectasis. Vascular intimal proliferation of the pulmonary vessels causes pulmonary hypertension and, eventually, cor pulmonale. Although the chest film may be normal, the finding of middle and lower lobe fibrosis with scattered 5 to 10 mm diameter cysts in a young woman is quite suggestive.

The barium enema characteristically reveals large pseudodiverticulae of the inferior border of the transverse colon early in the course of the disease. Loss of propulsive power with dilitation can occur at any level of the GI tract, duodenal dilatation may be a specific finding.

Molars are involved early with resorption of some portion of the alveolar socket (but without apparent loosening of the teeth). Jaw X-rays reveal widening of the periodontal space in as much as 30% of cases and when present may be pathognomonic of this disease (Oral Surg 6:483 '53).

Tapering or absorption of bony tips (fingers, clavicular tips) may be noted radiographically. Calcification at sites of repeated trauma is common.

SCURVY

Gerarde (Herbal, 1597):*

The gums are loosened, swolne and exulcerate; the mouth greviously stinking; the thighs and legs are withall verie often full of blewe spots, not unlike those that come of bruses; the face and rest of bodie is of times a pale color; and the feete are swolne, as in the dropsie.

* from WSC Copeman, Doctors and Disease in Tudor Times, Dawson, London, 1960.

"Scorbutic gooseflesh" is follicular hyperkeratosis with perifollicular purpura, which is both an early and characteristic finding in scurvy. Hairs in affected follicles become dry and fragile and at times are noted to stand straight away (perpendicular) from the skin surface (a diagnostic finding).

Gum changes do not occur in edentulous individuals.

Some X-ray findings in childhood are:
1. peripheral calcification in epiphyses (Wimberger's line or the halo sign)
2. dense distal metaphyseal line (Fraenkel's line)
3. rarefaction proximal to the distal metaphysis (Truemmerfeld zone)
4. fracture of the Truemmerfeld zone with impaction of the distal calcified fragments as a lateral jutting spur (Pelkan's spur).

(BACTEREMIC) SHOCK

Vacuolated polys in finger tip blood smears are indicative of septicemia (Krevans, JHH).

With shock +
1. urinary tract infection, suspect E. Coli
2. skin lesions, suspect pseudomonas (or staph)
3. nephrolithiasis, alkaline urine, and hospital exposure, suspect Proteus
4. prolonged in-hospital positive pressure respirator use, suspect klebsiella
5. indwelling venous catheter use, suspect staph

JHH survey (Fekety, 1966):

1. organisms: E Coli 46%, Klebsiella 26%, Pseudomonas 15%, Proteus 6%, other 7%

2. a priori antibiotic sensitivity
   a. (single therapy)
      polymixin   70%
      kanamycin   58%
      chloromycetin 47%

   b. (combined therapy)
      Kana + Poly + a Pen   92%
      Kana + Poly   86%
      Ceph + Chloro   79%
      Poly + Chloro + a Pen   85%
      Pen + Strep   58%

The growth of gram negative organisms appears to be limited by high glucose concentrations, which suggests a therapeutic adjunct (Robson, US Army).

ANTIBIOTIC THERAPY ALONE IS NOT ENOUGH!

With gram negative septicemia there is decreased cardiac output, altered peripheral resistance, and poor venous return. Saline infusion can be lifesaving (although 5 or more liters may be required before the arterial blood pressure increases).

When saline infusion raises the central venous pressure but urine output and arterial pressure remain poor, INOTROPY and VASO-DILITATION should be considered, i. e. isoproterenol or norepine-phrine + alpha blocker infusion. Digitalis is definitely worthwhile. Pharmacologic doses of steroids have both of these effects and are quite helpful.

Vasoconstriction is necessary when there is pooling of blood in peripheral beds. This is most common with gram positive coccal septicemia (particularly in the elderly). Intravascular hemolysis is a complication of gram positive coccal septicemia which should be anticipated.

SICKLE CELL DISEASE (See also "HEMOGLOBINOPATHIES")

Patients with sickle cell disease are prone to bacterial infections, particularly with the pneumococcus, during childhood. Decreased serum opsonizing activity may be involved in this diathesis (NEJM 279:459 '68). It is well to recall, however, that pulmonary thromboembolism is common in sicklers and may exactly mimic pneumonia on clinical grounds. In the evaluation of a pulmonary infiltrate in a sickler, the finding of blister cells, helmets, tricornered hats, and thorn cells in the peripheral blood smear (i. e. indications of a microangiopathic state) indicate pulmonary embolism rather than pneumonia as the process (JAMA 203:569 '68).

Transfusion increases blood viscosity. Therefore, exchange transfuse if more than one unit is required, lest a severe thrombotic episode be precipitated.

Systolic murmurs are audible in as much as 90% of individuals with sickle cell disease. Of these, about 70% are of maximal intensity in the mitral region. In addition to flow murmurs related to anemia per se, there appears to be a high incidence of papillary muscle dysfunction, perhaps because of obliteration of the end-arteriolar supply in this 'infarct-susceptible' region.

Sickle cell anemia may be quite difficult to differentiate from acute rheumatic fever on purely clinical grounds. The third heart sound is almost invariably present and at times is slightly prolonged, suggesting a diastolic rumble (Harvey, Georgetown). The electrocardiogram usually reveals ST and T changes, and in about 30% of cases, the PR interval is slightly prolonged. Examination for the eye signs of sickle cell disease at the bed side may be enlightening in the differential diagnosis.

The eye signs of sickle cell (SS) disease are (Arch Ophth 75: 353 '66):
1. venous tortuosity (47%)
2. small, refractile exudates (29%)
3. "sunburst" pigmentation (43%)
4. obliteration of arterioles and venules (49%)
5. angioid streaks (?)
6. conjunctival capillary segmentation (see "CONJUNCTIVA")

Doubled, distal flexion creases (separated by two or more epidermal ridges) of the fingers is said to be a good indication of sickle cell disease (Am J Dis Child 113:271 '67), although further verification of this cutaneous sign is at present necessary. Some degree of arachnodactyly is common in both sickle cell disease and sickle trait.

The dithionite (elution) slide test (Am J Clin Path 52:705 '69) is a rapid, qualitative test for the presence of hemoglobin S, which may be of value when the demands of time make the metabisulfite preparation unattractive.

SJÖGREN'S SYNDROME

The components of Sjögren's syndrome are the 'sicca' complex (keratoconjunctivitis sicca and xerostomia) plus rheumatoid arthritis or other connective tissue disorder (including latent disease reflected solely by serologic abnormalities).

An early manifestation is said to be an unusual distribution of dental caries, in which the incisors are involved (Oral Surg, Oral Med & Oral Path 21:34 '66).

Tearing with exposure to peeled onions is a simple screening test.

Because of xerostomia, an history of compulsive sour ball use is often obtained.

The diagnosis can be confirmed with assurance (and with safety) by the appearance of the parotids in rapid phase, serial isotope (technicium) scintillation scans of the head (Zeiger, NIH).

Renal tubular acidosis is a complication of Sjögren's syndrome, though it (rarely) also complicates other collagen vascular diseases without the sicca complex. Coincident lymphoma, sarcoma, and macroglobulinemia have also been reported.

SPLEEN

Splenomegaly is uncommon with metastatic disease to the liver (except when the primary is pancreas). Therefore, with deranged liver function tests and splenomegaly, widespread primary liver disease is more likely.

Galen:

We may not observe any hydrops when the spleen alone is considerably enlarged. When the liver is also involved, hydrops will appear (see "ALCOHOLISM" for ref.).

Splenomegaly is an invariable concomitant of malaria, which ranks first in causes of spontaneous splenic rupture throughout the world. Malaria also predisposes the spleen to traumatic rupture. Intense pain with pressure over the posterolateral axillary region of the 9th left intercostal space (Pagniello's sign) suggests incipient splenic rupture.

Kehr's sign (see "DIAPHRAGM") is a reliable physical indication of splenic rupture. Should ileus not preclude barium instillation through an N-G tube, the finding of increased distance between the stomach and the lateral abdominal wall in the left decubitus position is an X-ray sign of this catastrophe (Am J Roent 99:616 '67). Gastric displacement can at times be appreciated in the plain film. When loss of a psoas margin accompanies traumatic, splenic rupture, coincident renal rupture is to be ruled out.

STRAIGHT BACK SYNDROME (Dis Chest 39:437 '61)

The absence of the normal thoracic kyphosis in this condition leads to compression of the heart, particularly the right ventricular outflow tract, between the sternum and the vertebral column.

The disorder is asymptomatic; however, on physical examination there is an ejection-type murmur at the base suggesting pulmonic stenosis.

The electrocardiogram may show a shift of the terminal QRS forces to the right, inversion of the lateral precordial T waves, and occasional premature ventricular contractions.

The PA chest film will show a prominent pulmonary artery segment and there may be some suggestion of "left ventricular hypertrophy." The lateral view, however, is diagnostic.

Recognition of this syndrome may avoid a long, costly, and perhaps hazardous work-up to rule out a congenital cardiac defect or a significant valvular dysfunction.

STROKE

The electrocardiogram in cases of cerebro-vascular accident may mimic that of myocardial ischemia, being marked by altered and delayed ventricular repolarization (Lancet 2:429 '64). EKG changes are about 5 times more common with cerebral hemorrhage than with cerebral infarction, they are:

1. ST depression, T flatened or inverted
2. abnormally tall T waves, TU fusion

3. QT prolongation

4. (least common and most specific) wide and deep, sharp, symmetrical, inverted T waves

Bible (I Samuel 25:37-38):

... But it came to pass in the morning, when the wine was gone out of Nabal, and his wife had told him these things, that his heart died within him, and he became as a stone. And it came to pass about ten days after, that the Lord smote Nabal, that he died.

The internal carotid pulse can be palpated in its proximal portion through the pharynx (JAMA 152:321 '53).

"Drop attacks" (without disturbed consciousness) in later life are almost pathognomonic of vertebrobasilar artery insufficiency. Intermittent vertigo, without tinnitus or deafness, followed by visual changes (black spots, floaters, aura-like visual hallucinations, subjective visual distortion, field defects) are the most frequent early indications of vertebro-basilar insufficiency. Symptoms:

1. vertigo
2. visual acuity changes
3. ataxia
4. sensory hallucinations
5. episodic weakness
6. disturbed consciousness

are typically recurrent and vary in intensity and multiplicity with time. During the interim, patients appear well and are without objective signs of neurologic damage.

Amaurosis fugax in the elderly suggests carotid insufficiency as does unilateral glaucoma and unexplained unilateral ocular inflammation or blindness.

Urinary incontinence localizes the ischemic cerebral lesion to the parasaggital cortex.

Ice cold calorics (i.e. cold water in the ear) overcomes the supranuclear conjugate gaze paralysis which may occur with carotid occlusive disease but does not influence nuclear or infranuclear conjugate gaze palsy.

McKinney (Early Medieval Medicine, Johns Hopkins Press, 1937):

(Charles II of England). ... while shaving fell unconscious in his bed room. The following treatment was employed by the Royal physicians. A pint of blood was extracted from his right arm, then eight ounces from the left shoulder, next an emetic, two physics, and an enema consisting of 15 substances. Then his head was shaved and a blister raised on the scalp. To purge the brain a sneezing powder was given; then cowslip powder to strengthen it. Meanwhile more emetics; soothing drinks; and more bleeding; also a plaster of pitch and pigeon dung applied to the royal feet. Not to leave anything undone, the following substances were taken internally: melon seeds, manna, slippery elm, black cherry water, extract of the lily of the valley, peony, lavender, pearls dissolved in vinegar, gentian root, nutmeg, and finally 40 drops of extract of human skull. As a last resort bezoar stone was employed. But the Royal patient died.

TETRALOGY OF FALLOT (See also "CONGENITAL HEART DISEASE")

Cyanosis with a radiologically small heart implies for all practical purposes a diagnosis of tetralogy of Fallot (Taussig, JHH).

Large VSD and some degree of RV outflow obstruction are the essential anatomical features of Fallot's tetralogy. Variants may also have:

1. right aortic arch (about 20% of all cases)
2. persistent left superior vena cava
3. anomalies of the main pulmonary artery or its branches
4. absence of the pulmonic valve
5. aortic incompetence
6. anomalies of the coronary arteries

Squatting benefits cyanotic children by relieving postural hypotension, increasing aortic root pressure (leading to increased pulmonary blood flow), and decreasing the volume of poorly oxygenated blood returned from the legs. Initially, squatting is assumed effortlessly by children and is usually taken by parents as a form of play.

During the first year and a half, symptomatic children characteristically present with episodic hyperpnea, which occurs with effort (such as feeding or straining) and after a prolonged sleep. Cyanosis, tachycardia, and obtundation are related features of these episodes, which can be abolished by IV propranalol.

Left atrial enlargement is not a feature of this disorder, and its presence in a cardiac series suggests a mistaken primary diagnosis.

Notching of the lower left cardiac margin in a deep inspiratory chest film is present in about 25% of cases and indicates the presence of two functional ventricles, excluding a diagnosis of tricuspid atresia. Notching along the upper left cardiac margin at the level of the crista supraventricular defines the level of the RV outflow obstruction at the infundibulum.

The electrocardiogram reveals a complete vector reversal from V1 to V2 (i.e., predominant RV1 shifting to SV2, the Donzelot sign) in about 50% of all cases. RVH is present. Severe right axis deviation is not a feature, and accordingly, when the axis is greater than 150º, another diagnosis, such as transposition, should be entertained.

The electrocardiogram reveals the degree of RV outflow tract obstruction, for when stenosis is marked there is bidirectional shunting, left ventricular enlargement and prominence of the left precordial R wave. With pulmonic atresia, left precordial q waves appear, indicating left ventricular volume overload.

Some dermatoglyphic features of Fallot's tetralogy are:
1. 10 whorl pattern (Brit H J 26:524 '64)
2. unilateral right hypothenar loop (35-40% of males with tetralogy versus less than 10% of the "normal" male population)

Right-sided heart failure in patients with tetralogy may follow the development of pulmonic insufficiency or the apposition of the septal leaflet of the tricuspid valve against the ventricular septal defect, decreasing the RV runoff.

A feeling of heaviness in one leg is the earliest and perhaps the most reliable indication of thrombophlebitis, very often preceding the appearance of objective physical signs by 24 to 48 hours. At times an actual difference in the weight of a leg can be appreciated at the bedside when the subjective complaint is first made (O'Neal Humphries, JHH).

Deep vein thrombosis is easily missed on physical examination.
1. Palpation must be antero-posterior
2. When Homan's sign is negative, the test should be repeated while the patient is actively plantarflecting
3. When palpation and Homan's sign are not revealing, a blood pressure cuff should be applied to the thighs and inflated slowly until pain or 300 mm Hg is reached. A difference in pain threshold of greater than 30 mm Hg strongly suggests thrombophlebitis in the low threshold leg (this is a variant of the test reported in JAMA 155:1566 '54)

Thrombophlebitis associated with neoplasia is often in 'unusual' locations, such as the upper extremity. Pulmonary embolism is common in this situation.

Septic pelvic thrombophlebitis should be considered as a cause of FEVER in the postpartum period or after pelvic surgery when there are no localizing physical findings and no response to antibiotics (blood cultures are often positive in this situation). Anticoagulation is both diagnostic and therapeutic, having a definite antipyretic effect (NEJM 276:265 '67) (a similar temperature effect is not seen in cases of pelvic cellulitis or incisional wound infection).

THYROID - HYPERTHYROIDISM (See also "SERUM THYROID STATUS TEST")

When the liver is not palpable, the diagnosis of hyperthyroidism in a young individual is in doubt (E. Smith, JHH). Slight to moderate BSP retention is almost invariably present. The spleen is palpable in slightly less than 10% of cases.

The clinical features of hyperthyroidism may be masked by chlorpromazine use.

Although pretibial myxedema is a sign of advanced disease, malleolar region edema is an early feature, particularly in young women (Rev Lyon Med 16:323 '67). (Cardiac failure is not the cause.)

Plummer's lines, which occur most often on the ring fingers is a valuable diagnostic sign which should be carefully sought (Asper, JHH). Plummer (1937):

...the most striking change in appearance is produced by the irregular separation of the soft subcutaneous structures from beneath the surface of the nail, which makes cleaning of the nails difficult and allows dirt to accumulate under the nail, producing a black, irregular band.

The pulmonary artery segments are often quite prominent on chest films and may indeed suggest post-stenotic dilitation of the main pulmonary arteries. When high output failure is present, vascular congestion is florid, but cardiac size is not increased.

Palpitations occurring after the patient retires to bed is said to be a good clue of hyperthyroidism (Ann Int Med 69:1015 '68).

The systemic manifestations of hyperthyroidism may be lacking in elderly men in whom myopathy is the presenting feature (JC Harvey, JHH). In these cases there is marked weight loss, anorexia, weakness, emotional lability, and muscle wasting, particularly of the shoulder girdle, temporalis region, and of the small muscles of the hands.

Diarrhea is characteristic of hyperthyroidism, though it is rare. On the other hand, hyperthyroid patients are almost never constipated. An history of continued laxative use, therefore, makes the diagnosis of hyperthyroidism unlikely.

Apathetic hyperthyroidism may be difficult to differentiate from hypothyroidism. In both there is lid puffiness. With myxedema edema extends to the lid margin, but with hyperthyroidism it does not (Asper, JHH).

With hyperthyroidism there is marked epinephrine sensitivity. Topical application of a very dilute epinephrine solution to the conjunctiva results in marked mydriasis and at times even lid retraction. There is, too, brisk local reactivity to subcutaneous injection of serotonin or histamine, which is quantitatively greater than the norm (Acta Allerg 24:280 '69).

The lid lag of hyperthyroidism is often jerky and is best elicited with a rapid target. (A jerky lid lag can also be seen with extrapyramidal disease when it, possibly, represents a 'cogwheel' rigidity phenomenon.) Limitation of extraocular muscle motion occurs in dysthyroidism, independent of exophthalmos; unilateral limitation of upward gaze with diplopia is seen most often and follows lymphocyte infiltration of the vertical rectus muscles.

THYROID - HYPOTHYROIDISM

Dirty elbows in an otherwise clean patient are a reliable confirmatory sign of hypothyroidism (Berczeller, NYU).

The presence of axillary hair in hypothyroid patients mitigates against a pituitary causation.

Breast cancer is twice as common in hypothyroid women than in the general population.

Patients with myxedema are sensitive to pharmacologic agents. Particular care must be exercised when giving these individuals general anesthetics or cardiotonics.

CSF protein levels are often slightly elevated with hypothyroidism and do not of themselves indicate a CNS disorder.

Serum carotene is usually elevated with myxedema. Because of the simplicity of the lab assay, this test might well be included in the initial work-up.

Alveolar hypoventilation complicates myxedema. Respiratory insufficiency is a leading cause of death in these patients (Moser, Georgetown).

Osler (1914): (op. cit., see "GOUT")

Dyspnea is not a common accompaniment of ordinary tuberculosis. The greater part of one lung may be diseased and local trouble exists at the other apex without any shortness of breath. Even in paroxysms of very high fever, the respirations may not be much increased. Dyspnea occurs (a) with the rapid extension in both lungs of a bronchopneumonia; (b) with the occurrence of miliary tuberculosis; (c) sometimes with pneumothorax; (d) in old cases with much emphysema, and it may be associated with cyanosis; (e) and, lastly, in long standing cases, with contracted apices or great thickening of the pleura, the right heart is enlarged, and the dyspnea may be cardiac.

The temperature of consumptives is easily influenced by trivial causes which would not affect a normal person, such as mental excitement, exercise, constipation, etc. The patient is usually aware when fever is present and feels more comfortable with a temperature of 101°.

Loss of strength may be out of proportion and quite independent of weight loss.

Special examination should be made of the clavicular regions to see if one clavicle stands out more distinctly than the other, or if the spaces above it or below it are more marked. Defective expansion at one apex is an early and important sign...... standing behind the patient and placing the thumbs in the supraclavicular and the fingers in the infraclavicular spaces, one can judge accurately as to the relative mobility of the two sides. One of the earliest and most valuable signs is defective resonance upon and above the clavicle.

Renal tuberculosis should be considered when there is pyuria and routine urine cultures are negative. In women, renal tuberculosis is generally associated with an history of infertility.

Lower lobe pulmonary involvement can occur in diabetics.

Myoedema often accompanies active pulmonary tuberculosis.

Pleural involvement occurs early in the course of pulmonary tuberculosis and may present with pleural effusion, particularly in the young patient. When pleural effusion is associated with a strongly positive skin test but there are no signs of pulmonary or pleural disease, spinal tuberculosis should be ruled out.

A positive skin test does not of itself indicate active tuberculosis, although a febrile response with an intense skin reaction and local lymphadenitis is suggestive of activity.

A modest fever regularly occurring during the premenstrual phase of the cycle is often noted in women with pulmonary tuberculosis who are otherwise asymptomatic (JAMA 95:13 '30).

TUBEROUS SCLEROSIS (Bourneville's syndrome)

Tuberous sclerosis is one of the phakomatoses along with:
  neurofibromatosis
  Von Hippel-Lindau disease
  Sturge-Weber disease
  ataxia-telangiectasia

There is dominant inheritance with (apparently) variable penetrance. The family history may contain 'skip' generations.

> The classical triad is:
> adenoma sebaceum
> seizures
> mental retardation

The disorder is associated with a variety of soft tissue tumors (in addition to cutaneous fibroadenomas), renal hamartomas being, perhaps, the most common. Subungual fibromas are diagnostic. At times, cystic lesions can be palpated within the finger pads.

In the French literature, infantile flexor spasms are said to be the commonest early manifestation of tuberous sclerosis (Ann Ped 42: 770 '66).

The skull X-ray may reveal intracranial calcifications. In about half of all cases there is hyperostosis of the cranial vault, and the combination of these two X-ray findings in the skull is most suggestive of this diagnosis. Bone survey may also reveal polydactyly, periosteal irregularity in the metatarsals and metacarpals (Brit J Radiol 32:157 '59), and spina bifida.

Café au lait spots may be present, but there is no association with pheochromocytoma (as there is with neurofibromatosis). 'Shagreen' skin (a fine peau d'orange change) is common, particularly in the lumbosacral region.

Gray to cream color multinodular tumors of the retina (mulberry lesions) are diagnostic.

TULAREMIA

Tularemia is present along the eastern seaboard! About 1 to 2% of the rabbits and squirrels in Maryland are estimated to be infected. A definite history of animal exposure is not mandatory for the diagnosis in that it may be contracted from intermediate vector (insect) bites.

Without treatment there is a 10% mortality in the ulcero-oculo-glandular form and 60% mortality in the typhoido-pneumonic form. Therefore, TREAT AS SOON AS THIS DISEASE IS SUSPECTED (Streptomycin 2 gm/day).

Tularemia should be suspected when there is high fever and relative bradycardia during or close to the hunting season.

The infecting agent is a primary pulmonary pathogen. AEROSOL transmission occurs from animal to man, but, strangely enough, not from man to man; therefore, isolation is not necessary. Spread, however, is rapid from artificial media... this agent is TOO DANGEROUS TO CULTURE ARTIFICIALLY. The diagnosis confirmed by biopsy.

ULCERATIVE COLITIS (See also "COLITIC ARTHRITIS")

The causes of death in an autopsy series were (Arch Int Med 117:377 '66):

> 1. peritonitis   35%

2. neoplasia   11%
3. thrombosis 10%

i. e. an HYPERCOAGULABLE STATE may be present.

In a study of 9 patients (gastroent 54:76 '68), 7 with active disease and extensive colon involvement had increased factor VIII and thromboplastic generation acceleration, 5 had significant increase in fibrinogen level.

Toxic granulation, myeloid immaturity, and thrombocytosis are proportional to severity and can be evaluated from the peripheral smear when the patient presents (Am J Gastro 40:601 '63). Other hematologic changes are monocytosis (uncommon) and macrocytosis (which may indicate associated liver damage!).

Basophils accumulate in injured skin of ulcerative colitis patients, both erythema multiforme and erythema nodosum may occur. Cutaneous manifestations of ulcerative colitis are:

1. recurrent pyoderma gangrenosum, furunculosis
2. dermatitis herpetiformis-like eruption
3. neurodermatitis
4. aphthous stomatitis (papulonecrotic oral pyoderma may also occur)

WEGENER'S GRANULOMATOSIS

Wegener's granulomatosis is a triad of
respiratory tract involvement
diffuse vasculitis
glomerulitis

It is NOT the same as midline granuloma.

90% of cases have upper respiratory tract symptoms - epistaxes, rhinorrhea, septal and palatal lesions; 90% of cases have X-ray appreciable clouding of the air sinuses.

50% of cases have lower respiratory tract involvement - cough, pleurisy, hemoptysis.

40% have ocular manifestations - episcleritis, conjunctivitis, and (10%) proptosis.

50% have skin lesions - multiple purpuric or papulo-necrotic changes.

In addition, hematologic examination reveals:
1. anemia (90%)
2. leukocytosis (70%)
3. thrombocytosis (50%)
4. eosinophilia (40%)

WHIPPLE'S DISEASE

Whipple's disease is a chronic, relapsing disorder progressing to diarrhea, steatorrhea, and cachexia, which occurs 8 times more frequently in men than in women.

The clinical features are:
1. large joint arthritis (75%) (see "COLITIC ARTHRITIS")
2. serous cavity effusions
3. lymphadenopathy (50%)
4. intermittent fever (25%), night sweats
5. vague abdominal pains
6. episodic blurring of vision, flake-like vitreous opacifications, intraocular inflammation
7. tonsillitis (biopsy may be diagnostic)
8. skin hyperpigmentation (infrequent)

This disorder has a BACTERIAL etiology. Atypical organisms (pleomorphic gram negative cocci and rods, reverting to strep on hypertonic media) can be cultured from the blood or lymph nodes (see "ATYPICAL BACTERIAL ORGANISMS"). Chloramphenicol, erythromycin, and tetracycline are therapeutic.

The definitive diagnosis is made by rectal, gut, or (at times) tonsil biopsy, which discloses characteristic PAS positive material (representing bacterial cell walls or polysaccharide capsular material after lysosome digestion within macrophages).

WILSON'S DISEASE

Kayser-Fleisher rings are virtually always present in symptomatic cases, although a slit-lamp examination may be necessary for identification. They regress with adequate therapy.

A presumptive diagnosis of Wilson's disease is less likely when the serum uric acid is not decreased. Serum P is usually also decreased, and the urine is generally alkaline. In addition to uricosuria and phosphaturia, glycosuria and aminoaciduria are concomitants of renal tubular copper deposition.

Bone films often reveal vertebral osteochondritis or wedging. Osteoarthritic changes and bone fragmentation of the small joints of the hands and wrists may occur early (Brit J Radiol 32:805 '59). Some degree of dysarthria is present in most cases.

Transaminase levels seem to be the earliest indicators of hepatic involvement in this process. Serum copper levels do not differentiate homo- and hetero-zygotes. (Lancet 2:575 '67).

Early diagnosis is imperative. Liver biopsy with total copper concentration assay should be performed in all cases of unexplained cirrhosis at an early age and in any relatives of individuals with Wilson's disease who have elevated transaminase levels.

WOLFF-PARKINSON-WHITE SYNDROME

The electrocardiographic features are:
1. short PR interval
2. QRS duration prolongation
3. QRS slurring by a delta wave

In type A the delta wave is oriented anteriorly and to the right, in type B it is oriented to the left. Secondary T wave changes are present at rest or can be induced by exercise.

In 70% of cases there is no associated underlying heart disease. The male:female ratio is approximately 5:2. There appear to be associations between this syndrome and hypertrophic subaortic stenosis and Ebstein's anomaly.

About 10% of individuals with these EKG findings have recurrent arrhythmias. Paroxysmal atrial tachycardia is most common. Although atrial fibrillation accounts for about 10% of the observed arrhythmias, atrial flutter is only rarely found.

Paradoxical splitting secondary to early pulmonic valve closure may be detected on physical examination (Am H J 70:595 '65).

Delta waves may mimic or mask infarction Q waves. Whenever possible, WPW syndrome should be converted to normal sinus rhythm so that ventricular conduction can be studied. This can at times be accomplished with carotid massage, xylocaine, quinidine, or procaine amide. Arrhythmias can at times be converted with intravenous atropine which seems to facilitate 'normal' conduction through the A-V node.

Propranalol appears to be the agent of choice for prophylaxis against arrhythmias. Quinidine is a good choice when propranalol is contraindicated.

LABORATORY FINDINGS

Bernard:

THE SCIENCE OF LIFE IS A SUPERB AND DAZZINGLY LIGHTED HALL WHICH MAY BE REACHED ONLY BY PASSING THROUGH A LONG AND GHASTLY KITCHEN.

Pasteur:

WITHOUT LABORATORIES THE PHYSICIAN AND CHEMIST ARE SOLDIERS WITHOUT ARMS ON THE FIELD OF BATTLE.

ALDOSTERONE

The highest levels of aldosterone are found in cases of malignant hypertension. Elevations are also found with:

1. nephrosis
2. congestive heart failure
3. hepatic cirrhosis
4. shock
5. Conn's syndrome (primary aldosteronism)

Aldosterone secretion is greater in cases of cirrhosis than with congestive heart failure. With diuretic therapy in cirrhosis, K loss is a problem; conversely, only minimal K supplementation should be used in the initial phase of congestive heart failure, lest hyperkalemia be caused.

The major diagnostic aspects of Conn's syndrome are:
1. diastolic hypertension
2. failure to suppress aldosterone with salt loading
3. increased urinary K loss with salt loading
4. inability to stimulate plasma renin activity with volume depletion or BP lowering

Minor aspects are:
1. hypernatremia, hypokalemia
2. alkaline urine
3. reduced salivary Na/K ratio
4. decreased J-G index in renal biopsy

ANEMIA

The features of pyridoxine responsive anemia are:

1. peripheral smear reveals a dimorphic population of normocytic and normochromic cells; target cells and siderocytes may be present

2. the marrow is marked by erythroid hyperplasia; 10% have megaloblastic maturation; nucleated red cells have increased iron

3. the reticulocyte count is low; SERUM IRON AND IRON BINDING CAPACITY ARE HIGH DESPITE HYPOCHROMIA; indeed, iron overload may follow transfusion

4. hemoglobin electrophoresis is normal (the smear may initially suggest thallasemia)

5. the defect is in hemoglobin synthesis and is inherited as a sex-linked recessive or acquired with:
   a. rheumatoid arthritis
   b. primary carcinoma
   c. myeloproliferative disorders
   d. INH therapy, lead poisoning

6. there is no clinical evidence of pyridoxine deficiency (namely, glossitis, dermatitis, peripheral neuropathy)

Red cell aplasia is associated with thymoma and occasionally accompanies myasthenia. IF ANEMIA IS REFRACTORY TO STEROIDS, ANDROGENS, RIBOFLAVIN, OR SPLENECTOMY (and myeloproliferative changes have been excluded), RULE OUT THYMOMA, with a sternal splitting exploration if necessary.

The peripheral smear in cases of microangiopathic hemolytic anemia (of Brain and Dacie, also called non-immune hemolytic anemia or the hemolytic-uremic syndrome) reveals bizarre, distorted red cells (i.e. burrs, schizocytes, tricornered hats, etc.). The Coombs' test is negative. This disorder is seen with:

1. uremia and malignant hypertension, eclampsia
2. carcinomatosis (generally adrenocarcinoma. The mechanism may involve tumor thromboplastin production)
3. renal cortical necrosis
4. widespread collagen vasculitis
5. severe burns
6. thrombotic thrombocytopenic purpura
7. heart valve prosthesis (or severe rheumatic valvular distortion)
8. pulmonary embolization (see "SICKLE CELL ANEMIA")

ATYPICAL BACTERIAL FORMS (ABF)

ABF are protoplasts or L- forms; they are not the same as mycoplasma organisms. With proper culture techniques they revert to streptococci, although cultures must be stored for several weeks before evaluation.

ABF are related to Whipple's disease and cause some 25% of cases of primary atypical pneumonia. They are at times implicated in endocarditis in predisposed individuals receiving penicillin prophylaxis.

Arthralgias occur in less than 10% of cases of streptococcal endocarditis but are present in at least half of cases due to ABF's. The clinical picture of ABF-endocarditis is that of serum sickness with low grade, nightly, spiking fever. The ausculatory findings are less striking than with classical SBE. When ABF endocarditis is suspected, large volumes of blood (i.e. 10 cc per tube and 20-30 cc per bottle) should be cultured on thioglycollate. Gram stains from the culture media must not be heat fixed, or the ABF's will be destroyed.

BASOPHILIA

Basophils are increased in peripheral blood with:
1. myxedema
2. ulcerative colitis
3. myeloproliferative disorders, Hodgkin's disease
4. smallpox, chickenpox
5. after splenectomy
6. systemic mastocytosis

With ulcerative colitis, basophils increase in the intestine (during attacks) and accumulate in injured skin.

CALCIUM - HYPERCALCEMIA (See also "PARATHYROID")

The features of hypercalcemia are:
1. renal - thirst, polyuria, nephrolithiasis, azotemia (late)
2. gastrointestinal - nausea, anorexia, constipation
3. neuromuscular - weakness, hypotonia, lethargy, stupor (progressing to coma), paresthesias
4. cutaneous - pruritis
5. ocular - band keratopathy, burning, calcification
6. cardiac - EKG changes (QT prolongation), digitalis sensitivity

The causes of hypercalcemia are:
1. malignancy - metastatic to bone or with primary parathormone activity (in the separate case of breast cancer, a vitamin D-like sterol may be produced which increases GI uptake of calcium)
2. hyperparathyroidism
3. vitamin D intoxication
4. milk-alkali syndrome
5. sarcoidosis
6. renal tubular acidosis
7. osteoporosis of disuse (particularly with Paget's disease)
8. adventitious - use of cork stoppered test tubes
9. acute adrenal insufficiency
10. idiopathic (childhood)

The milk-alkali syndrome is characterized by hypercalcemia with CO_2 retention and high urine calcium.

With idiopathic hypercalcemia of childhood there is a characteristic "what me worry" facies, mental retardation, failure to thrive and an association with supravalvular aortic stenosis.

Eye calcifications are at times mistaken for an irritative conjunctivitis. Eye and other ectopic calcifications are associated with high calcium and high phosphorus.

The GI symptoms of hypercalcemia may resemble those of the peptic ulcer - hyperacidity group. In this situation, however, symptoms are intensified rather than ameliorated by a milk-antacid regimen.

COLD AGGLUTININS

The rapid screening test of Garrow enables the examiner to determine cold agglutinins of titer 1:64 or greater as a part of the immediate work-up of patients with pneumonia (Ann Int Med 70:701 '69). In this technique, a few drops of blood obtained by lancet puncture of a finger are mixed 1:1 with a citrate anticoagulant in a small test tube which is immersed in a cold water bath (0-4° C) for 30 seconds. Gross agglutination is noted by inspection.

ERYTHROPHAGOCYTOSIS

Leucocyte erythrophagocytosis in blood which has not been anticoagulated is ALWAYS an abnormal finding. It implies the presence of red cell coating with a complement activating antibody.

Erythrophagocytosis is seen with (Blood 7:592 '52):
1. bacterial infections - SBE, TB, typhoid
2. viral infections - hepatitis, mononucleosis
3. protozoal infections - malaria, trypansomiasis
4. certain chemical agents - naphthalene, potassium chlorate
5. Coombs' positive hemolytic anemias, transfusion reactions
6. leukemia

Erythrophagocytosis is not found with congenital spherocytic anemia or uncomplicated G-6 P D deficiency. Screening for this abnormality may be helpful in the initial work-up of hemolytic states.

Screening test (Birnholz): incubate venous blood with EDTA (or other) anticoagulant for 3 hours at room temperature, prepare a buffy coat smear. More than 5% poly erythrophagocytosis indicates antibody coating.

GLOBULINS

Panhyperglobulinemia (above 5 gm%) can be seen with:
1. chronic infection - TB, syphilis, SBE, leishmaniasis
LGV (Frei test or LGV titer should be done in all cases)
2. hepatic disease (particularly postnecrotic cirrhosis)
3. systemic lupus (but generally not polyarteritis)
4. histiocytosis X

Monoclonal hyperglobulinemia occurs with multiple myeloma and:
1. Waldenstrom's macroglobulinemia
2. leukemia, reticuloendothelial malignancies
3. certain carcinomas
4. polycythemia vera
5. renal tubular acidosis

A simple screening test for increased globulins is the formol gel reaction (2 drops of formalin in 2 cc serum, incubated for 3 hours at room temperature). The formation of a viscous or 'solid' gel is a positive test. Positive reactions also occur with substantial increase in fibrinogen level.

Essential cryoglobulinemia is complicated by progressive vascular sclerosis and collagen-vascular disease symptoms. Transient fragmentation of the blood column in conjunctival vessels after irrigation with cold water is pathognomonic of cryoglobulinemia and is performed in the office without morbidity

Agammaglobulinemia may be complicated by collagen-vascular disease symptomatology including a rheumatoid-like arthritis. Swiss-type agammaglobulinemia can be suspected from the lateral chest X-ray in the infant by absence of the typical thymic shadow. When absence of the thymic shadow is associated with hypocalcemia (in the infant) DiGeorge's syndrome is likely.

HAPTOGLOBIN

Haptoglobin levels are increased with extensive tissue necrosis or widespread metastatic disease.

The triad of   FEVER
ANEMIA
INCREASED HAPTOGLOBIN

has been suggested as an early indication of hypernephroma, preceding the more typical flank pain and hematuria (Ann Int Med 68:613 '68).

HYDROXYPROLINE

Hydroxyproline excretion is related to bone collagen metabolism. A normal urinary hydroxyproline determination in a patient with elevated alkaline phosphatase indicates hepatic disease.

Increased urinary hydroxyproline is found with (Arch Int Med 118:565 '66):
1. Paget's disease of bone
2. hyperthyroidism
3. hyperparathyroidism
4. psoriasis
5. extensive burns
6. Marfan's syndrome, Hurler's disease
7. malignancies with osseous involvement (minimal increase with osteolytic lesions, maximal elevation with osteoblastic changes).

LATEX FIXATION

Latex fixation may be positive in cases of rheumatoid arthritis, subacute bacterial endocarditis, and:

1. sarcoidosis
2. myeloproliferative disorders
3. chronic infectious processes, including syphilis
4. hepatitis, cirrhosis
5. thyroid disease
6. some malignancies

MACROCYTOSIS

Non-megaloblastic causes of macrocytosis are:
1. hypothyroidism
2. radiation exposure
3. multiple myeloma (in some cases there has been clear cut, coexistent pernicious anemia, though this is not always the cause of macrocytosis in this disease)
4. multiple sclerosis (without anemia, macrocytosis in this case indicates disease activity, Acta Med Scand 44:81 '68)

MAGNESIUM

Hypomagnesemia (Am J Med 41:645 '66) may follow prolonged parenteral fluid therapy with magnesium-free fluids. It is also seen with:

excessive urinary magnesium loss
primary aldosteronism, renal tubular acidosis, diuretic phase of renal tubular necrosis, prolonged diuretic therapy, inappropriate ADH secretion (including acute porphyria)

redistribution
hypoparathyroidism (including post-parathyroidectomy), hypercalcemia, pancreatitis, insulin therapy of ketoacidosis

decreased GI uptake or excessive GI loss
vomiting, diarrhea, prolonged therapeutic GI aspiration, chronic alcoholism, hepatic cirrhosis

The signs of hypomagnesemia are:
1. calcium resistant tetany
2. fasciculations
3. twitching
4. hyperactive reflexes
5. Chvostek sign (but not Trousseau's sign)
6. choreoathetoid movements
7. personality change, delirium, confusion
8. sweating, fever, tachycardia

Hypermagnesemia usually has an iatrogenic cause, particularly in the setting of chronic renal disease. It can also be seen with:
1. extensive tissue necrosis
2. Addison's disease
3. diabetic acidosis

PLATELETS

When platelets are significantly increased, serum acid phosphatase will be elevated.

Thrombocytosis and (paradoxical) hemorrhagic diathesis indicates a myeloproliferative disorder (the platelets are generally large and bizarre on smear).

Large platelets may be found with:
1. pseudoxanthoma elasticum
2. Marfan's syndrome
3. Hunter's syndrome, Hurler's syndrome
4. Ehlers-Danlos syndrome
5. osteogenesis imperfecta

Thrombocytosis occurs transiently post-operatively.

PLEURAL FLUID

Pleural fluid glucose greater than 100 excludes the diagnosis of pleural tuberculosis. Pleural fluid glucose is decreased slightly with bacterial infections and moderately with tuberculosis; it is extremely low, however, with collagen vascular diseases and normal when the effusion is secondary to infarction or congestive heart failure.

Significance of lymphocytosis (Ann Int Med 66:972 '67):

...a cellular effusion characterized by severe lymphocytosis and absence of basophilic mesothelial cells should be considered tuberculous or, less frequently, neoplastic until proved otherwise.

With tuberculosis, lymphs are mature, with lymphoma they are abnormal. When neoplasia is present, bizarre cells can be identified in at least 2/3 of cases. Basophilic and degenerated mesothelial cells are abundant when the effusion is due to congestive failure or pulmonary infarction.

Pleural effusion is often unilateral, confined to the right side, when due to early left-sided cardiac decompensation, probably because of the right lung venous drainage into the azogous system. Conversely, unilateral, left pleural effusion is unlikely to be due to congestive heart failure.

SEDIMENTATION RATE

When cryoglobulins are present, the sed rate is temperature dependent. An elevated sed rate may be masked at room temperature in this situation (indeed it may be particularly low) and should be performed at 37⁰ when there is doubt.

The sed rate is often decreased when there is congestive heart failure (perhaps because of decreased fibrinogen levels, Brit H J 24: 180 '62). In this case, the sed rate is not a reliable indicator of concomitant infection.

The sed rate is minimally elevated during oral contraceptive use (B M J 3:214 '67).

SEROLOGIC TESTS FOR SYPHILIS

Biologic false positive tests occur transiently with:
1. malaria
2. small pox vaccination
3. infectious mononucleosis
4. infectious hepatitis
5. lymphogranuloma venereum
6. mycoplasma pneumonias
7. Weil's disease
8. (rarely with) SBE, mumps

Chronic false positive reactions are 3 times commoner in women, of whom some 25% will develop a collagen vascular disorder at some time, other causes are:
1. leprosy
2. dysgammaglobulinemia

3. chronic liver disease
4. Hashimoto's thyroiditis

SERUM

Green colored serum is said to be present in some 12% of patients with rheumatoid arthritis, usually of many years duration (Ann Rheum Dis 27:151 '68). A greenish cast to the serum is also present in some women during pregnancy and with oral contraceptive use.

SPUTUM

The presence of large numbers of eosinophils in sputum is suggestive (but not pathognomonic) of asthma. With asthma, however, Curschmann spirals can often be seen in the sputum with a hand lens, appearing as minute yellow white spheres or wavy threads (up to a centimeter in length) with fine projecting raylike fibrils.

Mucopurulent sputum with predominance of mononuclear cells (rather than polys) indicates a rickettsial or viral infection. With varicella pneumonia, intranuclear inclusions can be seen in the mononuclear cells.

The sputa of bronchiectasis and lung abscess are increased in volume and exceedingly putrid in odor.

Rusty sputum is the classical appearance of lobar, pneumococcal pneumonia. Other color changes are:

1. dark green - pseudomonas infection
2. "egg yolk" - jaundice
3. bright yellow green - hepatic abscess ruptured into lung (bilious sputum)
4. "anchovy sauce" - hepatic amebic abscess ruptured into lung

Blood streaked sputum with large numbers of iron containing macrophages ("siderocytes") are characteristic of pulmonary hemosiderosis and of Goodpasture's syndrome. Siderocytes are regularly found in the sputum of patients with tight mitral stenosis.

SERUM THYROID STATUS TESTS

Serum free thyroxine is a valid test of thyroid status. PBI is influenced by exogenous iodine contamination. Both PBI and T4 measure total thyroxine (99+% bound) and are, therefore, influenced by binding protein changes.

Thyroxine binding globulin (TBG) is decreased (PBI and T4 falsely low) with:

1. androgen or anabolic steroid use
2. ACTH, glucocorticoid use (but not with Cushing's disease)
3. nephrosis
4. salicylate use
5. congenital TBG abnormality

It is increased (falsely high PBI, T4) with:
1. estrogen effect (pregnancy, oral contraceptive use, hydatiform mole, cirrhosis in males)

104

2. acute hepatitis
3. trilafon therapy
4. idiopathic TBG increase

TBG binding capacity is altered with dilantin therapy, the PBI is falsely low but the T4 (by Murphy-Pattee) is high.

Bard Intracaths contain iodine. PBI determinations will be adventitiously elevated in individuals who have such intravenous tubing in place for several hours (i. e. moribund, hospitalized patients) (Arch Int Med 123:587 '69).

URINE

Color:

black   alcaptonuria
                   carbolic acid or lysol intoxication
                   melanotic tumors
                   methemoglobinuria

blue-green   methylene blue ingestion

red-orange   rhubarb ingestion
                   cascara use (alkaline pH)
                   presence of bile pigments

pink-red   phenolphthalein use (alkaline pH)

red   hemoglobinuria
                   porphyrinuria
                   anthrocyanuria (excessive beet ingestion)
                   aniline dye (at times in sweets) ingestion
                   azo dye (as in azo-gantrisin) preparations
                   phenylindanedione (and related anticoagulant) use

The ferric chloride test:

green   phenylketonuria (PKU) (evanescent)
                   histidemia (persistent)
                   maple syrup urine disease
                   tyrosinuria

gray-green   chlorpromazine use

purple   salicylate ingestion
                   ketones (as with diabetic ketoacidosis)

Hemoglobinuria:

1. Approximately 25% of free hemoglobin is excreted by the kidney, the remainder is converted to bilirubin products or bound to albumin forming methemalbumin, which is excreted slowly. Increased bilirubin distinguishes this state from myoglobinuria (excretion of myoglobin, which is released after crush injury or extensive muscle necrosis, is primarily renal) and from hematuria with hemolysis in the urine (hemoglobinemia will be lacking in this state as well).

2. March hemoglobinuria transiently follows upright exercise and does not indicate serious underlying disease. The mechanism appears to be traumatic fragmentation of red cells traversing the capillary beds of the soles of the feet; correction is by providing a spongy, soft shoe (Sem Hematol 6:150 '69).

a. March hemoglobinuria does not occur in south-east Asia (despite the prevalence of malaria which increases red cell fragility), because of the soft, marshy terrain. It may, however, be more frequently noted in soldiers with malaria returning to this country with our paved road system.

3. Hemosiderinuria regularly occurs with the presence of heme pigments in the serum. In the absence of hemochromatosis, this finding should indicate intravascular hemolysis.

4. In paroxysmal nocturnal hemoglobinuria there is activation of the complement-hemolysis system during sleep (independent of the time of day the patient sleeps). The molecular defect is uncertain but appears to involve a red cell acetyl cholinesterase deficiency. Patients are prone to thrombotic episodes, which can at times be managed with coumarin type anticoagulants (heparin appears to accelerate hemolysis). Red cells from patients with this disorder readily lyse when incubated at 37° in acidified (pH 6.5) serum (Ham test).

ABDOMINAL X-RAY

Increased hepatic density suggests hemochromatosis (this sign can be appreciated from chest films). Increased hepatic density is also seen in the rare patient who underwent thoratrast contrast studies in years past.

Egg shell calcification in a suprarenal or para-aortic location can be seen with pheochromocytoma, as well as with an arterial aneurysm.

The right inferior margin of the liver can almost always be delineated in the normal study. Its absence is a reliable, early sign of intra-abdominal fluid accumulation. (Radiol 88:51 '67).

In the great majority of upright films in cases of adynamic ileus, there are NO FLUID LEVELS; in the remainder, short fluid levels are present. Long fluid levels suggests mechanical obstruction. Squaring of loop segments is also characteristic of adynamic ileus.

Air under diaphragms WITHOUT PAIN can occur with pneumatosis cystoides intestinalis. The etiology of this condition may be alveolar rupture with dissection of air through the mediastinum and retroperitoneum to the mesenteric vascular network causing multiple, subserosal gas filled cysts, which may rupture suddenly. Sealed perforation in elderly individuals may be painless. Interposition of the hepatic flexure between the liver and diaphragm (Chilaiditi's syndrome) is another cause of painless air under the (Rt) diaphragm, although in some of these cases there is nausea, vomiting, and ill-defined abdominal discomfort.

Large pneumoperitoneum almost NEVER occurs with acute appendicitis; when present colon rupture or perforated ulcer are more likely.

Cecal distension is without localizing value since it may occur with distal colon obstruction in the absence of X-ray evidence of transverse colon abnormality. A long segment of air in the colon wall in an acute abdominal problem suggests an ischemic event, when it usually presents in the cecum or sigmoid region. Colonic intramural air tracking is also seen with diverticulitis and granulomatous colitis.

Foci of metastatic colon and ovarian carcinoma in the liver frequently calcify. Since the primary site may be "silent," work-up of this finding must include proctosigmoidoscopy, barium enema, and GYN consultation. Calcifications in these cases are generally few and large. Other causes of this type of calcification are hepatic artery aneurysm, hydatid cysts, healed hepatic abscesses, and hepar lobatum.

Multiple, small hepatic calcifications are seen with granulomatous disease, particularly tuberculosis. Intrahepatic phleboliths are multiple small calcifications with a dense peripheral margin. Their presence is a rare, but definite, sign of hemangioma (as they are in any other location in the body).

Intussusception after age 3 or 4 implies an underlying bowel lesion. In youth, hamartoma is to be considered; with increasing age, cancer becomes a more likely cause.

Fine, mottled calcification in an upper quadrant mass during childhood suggests, first, neuroblastoma or hepatoblastoma.

With a right lower quadrant calcification in a young person, think first of appendicolith.

An enlarging spleen will displace the stomach posteriorly (as will an enlarged left hepatic lobe); a pancreatic mass will displace it anteriorly.

BONE X-RAYS

Dense and diffuse central vertebral interspace calcification is pathognomonic of ochronosis.

Foot X-rays may indicate diabetes by the following:
1.   localized or diffuse osteoporosis without history of disuse
2.   arterial calcification
3.   juxta-articular, crescentic cortical bone defects

Sickle cell disease is indicated in vertebral column X-rays by (Am J Roent 104:838 '68):

1.   coarse osseous trabeculation and demineralization (the posterior ribs may also be involved)

2.   "cod-fish" vertebrae - symmetrical depressions of adjacent upper and lower central end plates without change in rectangular vertebral body shape, most easily noted in the lower thoracic segments (this change follows vertebral body osteoporosis in children and osteomalacia in adults, it is noted with other hemoglobinopathies and with homocysteinuria).

3.   anterior vertebral vascular notching (the Riggs-Rockett sign), which can be seen in lateral chest films (although this finding is "normal" in infants, it is seen in less than 10% of routine films of children 3 to 6 years of age as compared with almost 50% of sicklers).

Pathological fractures occur with chronic osteomyelitis, primary bone tumors and cysts, marble bone disease, and:

1. osteoporosis
2. osteomalacia (adult rickets)
3. metastatic disease
4. radiation therapy and other causes of aseptic necrosis

Multiple vertical rib fractures suggest Cushing's Disease (in the absence of an history of radiation therapy).

Periosteal reaction involving the anterior or lateral borders of the vertebral bodies suggests Hodgkin's disease (Brit J Radiol 40: 939 '67).

Accentuation of kyphosis of the upper thoracic spine (with compensatory increased lordosis of the lumbar region) and appositional new bone formation along the anterior aspect of the thoracic vertebrae (which can be seen in lateral chest films) is characteristic of acromegaly (A J Roent 86:321 '61).

Long bone X-rays should be included in the work-up of children suspected of having a leukemia. The commonest early sign is a narrow, transverse radiolucent band beneath the metaphysis (this resembles the Truemmerfeld zone of scurvy, but the metaphysis is otherwise preserved). Although this sign is not specific early in life, after age 2, it is most indicative of leukemia. Later X-ray findings are osteolytic defects and periosteal elevation.

In evaluating lytic vertebral lesions, a valuable differential point is that myeloma is much less likely to destroy a pedicle than is metastatic cancer (AJR 80:817 '58).

A solitary, completely sclerotic vertebral body ("ivory vertebra") is seen with Paget's disease of bone, Hodgkin's disease, and osteoblastic metastasis. With Paget's there is, characteristically, increase in vertebral body size and with Hodgkin's, signs of lymphadenopathy may be found elsewhere on the film. The commonest osteoblastic origins are prostate and breast.

Dense metaphyseal bands are seen with intoxication with lead (or lead's radiographic imitators phosphorus and bismuth). Apropos, inhalation of tetra ethyl lead in petroleum will effectively raise blood lead levels. The fibular head is often first involved. Multiple metaphyseal bands are also seen with with vitamin D intoxication (and at least one case of idiopathic hypercalcemia of childhood, which may have a similar biochemical basis, Phister, MGH). The picture is imitated by metaphyseal bone proliferation following treatment of rickets.

When thoracic scoliosis is present in a boy, or when it is convex to the left in a young girl, underlying bone disease is to be ruled out. Neurofibromatosis, perhaps, leads the differential diagnosis.

A focally thickened skull table raises consideration of:
1. meningioma
2. fibrous dysplasia
3. Paget's disease

Soft tissue lesions of the scalp, braided hair, residual EEG paste, et cetera must not be mistaken for true hyperostosis.

CHEST X-RAY

A long air esophagram is indicative of <u>scleroderma</u>.

Asymmetrical pulmonary artery segment enlargement with prominence of the left side occurs with pulmonic stenosis, symmetrical enlargement occurs with intracardiac left to right shunts.

When the honey-comb pattern of scleroderma, sarcoidosis, etc. is associated with an history of recurrent pneumothorax, histiocytosis X should be considered in the differential diagnosis (NEJM 270:73 '64). When pneumothorax accompanies pulmonary fibrosis, history of any remote occupational dust exposure should be sought, especially to bauxite (aluminum oxide, Shaver's disease) and to diatomaceous earth.

A single fused sternum in an infant frequently indicates the presence of a congenital heart defect.

Left atrial enlargement is indicated in the plain lateral chest film by posterior displacement of the left bronchial tree (Radiol 93: 279 '69).

Erosion of the superior aspects of the posterior ribs may occur with the collagen vascular diseases (including rheumatoid arthritis) (Am J Roent 106:491 '69).

Calcifications:

1. lymph node egg shell - silicotuberculosis
2. mediastinal egg shell - thymoma, <u>dermoid</u>, substernal thyroid adenoma, hydatid cyst (these lesions <u>must</u> be differentiated from a superimposed aortic aneurysm with advential calcification)
3. crescentic - fungus ball
4. pop-corn - hamartoma
5. pleural - asbestosis (usually with associated linear fibrosis)

The commonest intracardiac calcification in adult American males is coronary artery calcification. In some instances, the blatant appearance of calcified 'tubes' may be noted, although the usual situation is that of discrete flecks.

Loss of the normally curved diaphragmatic contour (in the absence of pleural effusion) is seen with Hodgkin's disease and represents subpleural plaque formation. Kerley B lines indicate pulmonary infiltration. Subpleural metastases are the rule in sarcoma, although they are generally not appreciated in plain chest films.

Increased distance from the upper border of the diaphragm to the stomach bubble is seen with subpulmonic effusion, which is particularly common with nephrosis.

Lower rib ends in men are generally concave, in women pointed, convex, thus providing a convenient though not terribly reliable, sex index on the chest film. Another such indication is rectangular costochondral calcification in the male and triangular (pointed) calcification in the female (Janower, MGH).

When there is fracture of the first two or three ribs, look carefully for pneumomediastinum: this dyad indicates a bronchial tear.

Malignant pulmonary lesions rarely cross the tough, fibrous tissue of a major fissure. When this boundary is not respected, an inflammatory etiology is suggested. Peripheral, versus central, necrosis in a mass lesion is said to indicate malignancy rather than granulomatous disease with some degree of certainty.

ELECTROCARDIOGRAPHY

LEFT AXIS DEVIATION (mean frontal plane QRS direction at least -30°) is a conduction defect involving a peripheral radicle of the left bundle. It is not related to cardiac position. (See Am H J 72: 391 '66). It is seen with:

1. 'infiltrative' cardiomyopathies - amyloidosis, scleroderma, myocarditis, hemochromatosis, etc.

2. ischemic heart disease with fibrosis (when left axis deviation accompanies left ventricular hypertrophy it indicates fibrosis disrupting left bundle integrity)

3. localized disease of conductive tissue (in otherwise entirely normal individuals)

4. congenital heart disease in which the normal situation of the left bundle is disturbed - e.g. endocardial cushion defects and single ventricle.

Peri-infarction block (Grant) is that situation in which there is left axis deviation with an initial .04 sec QRS vector shifted to the right (pointing away, one might say, from an antero-lateral infarction).

Pseudo-left axis deviation occurs with chronic obstructive lung disease and is characterized by:

1. inferiorly directed P vector
2. low QRS voltage
3. terminal QRS vector shifted superiorly and to the right (s wave present in lead I)
4. terminal R of AVR greater or equal to terminal R of AVL

When a person in normal sinus rhythm is noted to have a frontal plane P wave axis beyond +90°, DEXTROCARDIA is present.

Transient or permanent complete A-V block is likely to develop in individuals with a complete block of one bundle and a partial block of the other (i.e. bilateral bundle branch block). Right bundle branch block and left axis deviation or right or left bundle branch blocks and prolonged PR interval are examples of the at-risk state. Indeed, the RBBB+LAD constellation was found in about 60% of persons with Stokes-Adams attacks in one series (Circ 37:429 '68). The occurrence of any of these EKG patterns with an history of episodic syncope is an indication for placement of a demand pacemaker.

The sudden appearance of tall peaked precordial T waves (or sharp reversal of inverted T's)
                without ST segment change
                and associated with clinical deterioration after cardiac surgery or myocardial infarction

indicates acute hemopericardium (Circ 25:780 '62).

Duchenne's progressive muscular dystrophy should be suspected from the following distinctive EKG picture:

1. abnormally tall right precordial R waves, with
2. deep (but not prolonged) q waves in limb and/or lateral precordial leads

Localized postero-lateral fibrosis is causative (Am J M 42: 180 '67).

An isolated upright T wave in V1 in children less than 10 years of age with an otherwise normal electrocardiogram (e. g. no T inversion V5, V6) implies right ventricular enlargement (J Ped 74:413 '69). This EKG finding, for example, in association with a systolic murmur indicates pulmonic stenosis.

Ventricular premature beats with the QS or Qr pattern in the inferior limb leads (in the absence of myocardial ischemia) are said to be 'diagnostic' of pulmonary embolization (Jap H J 11:195 '70).

DRUGS

Hippocrates:

TO DO NOTHING IS SOMETIMES A GOOD REMEDY.

Moliere (Le Malade Imaginaire):

MOST MEN DIE OF THEIR REMEDIES AND NOT OF THEIR ILLNESSES.

ARSENIC

Arsenic poisoning still occurs in this country either accidentally through exposure to arsenical pesticides or intentionally, particularly in the lower socioeconomic stratum of the rural population.

Many of the features of acute arsenical toxicity are explained by the potent capillary poisoning of this agent, i. e.

1. cutaneous - pruritis, palmar (and plantar) erythema localized edema (particularly face and eyelids)

2. gastrointestinal - acute dysphagia, acute abdominal pain, nausea, vomiting, dysenteric symptoms

3. renal - acute tubular necrosis

4. circulatory - acute depletion of effective blood volume with thirst, oliguria, progressing to anuria and shock

5. cardiac - depressed myocardial contractility, profound ischemic type ST and T EKG changes

The clinical features of chronic arsenic poisoning are:

1. general - weakness, lethargy, hyperhidrosis, excessive salivation, excessive lacrimation, alopecia

2. cutaneous - hyperpigmentation (generalized or reticulated and confined to the shoulders and back), palmar keratoses, eczematous eruptions, nail signs

3. gastrointestinal - diarrhea, hepatic failure

4. hematologic - bone marrow depression

5. mucosal - conjunctival engorgement, nasal mucosal inflammation, ulcerative stomatitis

6. neurological - depressed, slow mentation, peripheral neuritis (resembling that of lead poisoning but affecting the legs more than the arms)

Arsenic poisoning may be suspected with:
1. persistent coryza, particularly "out of season"
2. peripheral neuritis, herpes zoster infections
3. "garlic" perspiration
4. generalized hyperpigmentation, palmar keratoses

CHLORPROMAZINE (PHENOTHIAZINE DRUGS)

Chlorpromazine has an epinephrine blocking effect and may precipitate heart failure when used in a patient with only borderline cardiac compensation.

1 to 2% of individuals taking these drugs more than 2 weeks will have jaundice at some time. Of these, 70% respond promptly to drug removal; in others it may persist, but biliary cirrhosis is an unlikely complication.

Although the serum is clear, chemical determinations reveal elevations of cholesterol, lipids, and phospholipids. An obstructive pattern is demonstrated by laboratory tests.

These agents lower the convulsive threshold. When used in epileptics, it may be necessary to increase the analeptic drug dose. Overdose may cause oculogyric crisis (conversely, oculogyric crisis should immediately suggest phenothiazine overdose).

Chlorpromazine decreases bronchial reflexes. It should, therefore, not be used for sedation during acute asthmatic episodes.

Use of chlorpromazine is related to QT prolongation on the electrocardiogram. It should, therefore, be avoided in individuals with ectopic cardiac beats.

Chlorpromazine (also methyldopa and isoniazid) may stimulate production of anti-nuclear factor, causing a false impression of collagen vascular disease in some instances (Acta Med Scand 157:67 '70).

COCONUTS

Coconut milk is sterile and very nearly isotonic, though it is high in potassium (about 55 meq/L.) It can be used as a rehydrating fluid for children with diarrhea and can also be infused directly from the nut (through a blood infusion filter apparatus) in emergencies when more conventional IV fluids are lacking and coconuts are on hand. (See Lancet 1:1278 '67, 2:968 '66).

COUMARIN DRUGS

Coumarin inhibits metabolism of tolbutamide and dilantin; intoxication with these agents may be a problem when combined therapy with coumarin is begun.

The metabolism of coumarin is enhanced by sedatives, especially the barbiturates and chloral hydrate. Excessive amounts of the anticoagulant may be necessary in the hospitalized patient who is receiving concomitant sedation. The dose should be readjusted before discharge if possible.

Relative coumarin resistance is seen (in addition to concomitant sedative use) (Am J M 42:620 '67):

1. during lactation
2. with oral contraceptive use
3. with pulmonary infarction
4. with hyperlipidemia
5. as a familial defect

Formation of tumor metastases may involve coagulation with the localization of tumor cells within the forming thrombus. Coumarin anticoagulation at the time of surgery may reduce the liklihood of metastasis from tumor dissemination. Similarly, continued anticoagulation may be a useful adjunct in all patients with malignancies (see Cancer 15:276 '62).

DIGITALIS

Withering (1785):

The foxglove when given in very large and quickly repeated doses, occassions sickness, vomiting, purging, giddiness, confused vision, objects appearing green or yellow; increased secretion of urine with frequent motions to part with it, and sometimes inability to restrain it; · slow pulse, even as slow as 35 in a minute, cold sweats, convulsions, syncope, death. (I am doubtful whether it does not sometimes excite a copious flow of saliva.)

Specific therapy of digitalis toxicity with binding antibodies appears to be a promising mode of treatment from preliminary animal studies (Schmidt & Butler, Columbia).

Hypomagnesemia facilitates digitalis toxicity as does hypokalemia. Serum magnesium should be determined in cases of toxicity in clinical situations where it might be reduced (see MAGNESIUM) and when the initial potassium determination is normal.

The tensilon test   useful in the diagnosis of digitalis toxicity (Pitt, JHH). 5 mg are given IV, effects are noted within 2 minutes and persist from 10 to 15 minutes:

1. with NO or INADEQUATE digitalis, heart rate will not slow more than 30 beats per minute

2. with DIGITALIS TOXICITY there is significant slowing with increased incidence of premature ventricular contractions

This test is dangerous in that a serious arrhythmia may occur when the heart rate is slowed. The test should be preceded by various physiologic vagal stimuli and performed when these procedures do not cause slowing with the appearance of irritable foci, in itself, a positive test.

DIPHENYLHYDANTOIN (Dilantin)

Dilantin intoxication may present in chronic users as drowsiness, behavioral change, nystagmus, blurred vision, or diplopia. The possibility of intoxication should be considered in those situations in which drug metabolism is retarded but the regular dose has been maintained, i.e. hepatic disease, intercurrent infection, late initiation of barbiturate therapy.

An uncommon complication of chronic dilantin therapy is a macrocytic (megaloblastic) anemia, which is secondary to small bowel folate malabsorption. Although naturally occurring folates (polyglutamates) are not absorbed, folic acid (monoglutamate) is therapeutic (Lancet 2:528 '68). When supplemental folic acid is used, aggravation of the seizure disorder may occur.

Intravenous dilantin (administered SLOWLY) can be used to control ventricular myocardial irritability and to convert digitoxic supraventricular arrhythmias. Pretreatment with dilantin in individuals with ventricular irritability may enable a fully inotropic dose of digitalis to be given without increasing the danger of a catastrophic ventricular arrhythmia (Clin Res 15:206 '67, abstr).

DRUG OVERDOSE

Miosis is seen in coma from overdose with a variety of sedative agents. Pupillary dilatation with topical application of Nalline is a valid differential sign of narcotic use.

Fever and delirium in a narcotics addict may be secondary to a superimposed septic process, but the possibility that barbiturate withdrawal is causative (or contributory) must be kept in mind. Likewise, nystagmus in a narcotics addict suggests concomitant barbiturate overdose.

Trophoneurotic blisters may follow overdose with any sedative, although when associated with barbiturate overdose, breakdown to decubitus ulceration is rapid.

Aspiration pneumonitis is frequent with narcotics overdose, largely because of the efforts of the patient's friends to revive him by forcing oral fluids (particularly milk, which is often used as a 'specific' antidote).

Rapidly progressive and fatal pulmonary edema may follow unsterile intravenous drug use, usually narcotics and less often amphetamines (the association of opiate use and pulmonary edema was noted by Osler before the turn of the century, Montr Gen Hosp Rep 1:291 1880). Physical signs may be minimal but the chest X-ray is florid. Chest films are mandatory at 12 and 24 hours after admission, regardless of the physical findings. The etiology of this complication is uncertain, although it resembles endotoxin effects in experimental studies.

Narcotics withdrawal symptoms are often exaggerated by addicts, who, perhaps, feel some need to convince themselves of the potency of the mixtures they have been purchasing. Rhinorrhea is a valid sign of withdrawal, and in its absence it may be well to temporize with parenteral vitamin C or B-C mixtures which mimic demerol and methadone in causing local irritation at the injection site.

In addition to forearm "tracks," another cutaneous sign of narcotics use is a colarette of cigarette burns about the upper chest, which occur with drug torpor while smoking.

Pupil dilatation is seen with amphetamine, LSD and STP use. With amphetamines there is facial flushing, irritability, hyperflexia, restlessness and occasionally paranoid ideation. STP is an amphetamine derative with atropine-like side effects.

Atropine-like effects are seen with suicidal attempts with proprietary sleep medicaments (i. e. Benadryl overdose), with nutmeg intoxication (NEJM 269:36 '63), and with stramonium intoxication following use of tea or cigarettes prepared from 'Asthmador', a proprietary spasmolytic for asthma (Ann Int Med 68:174 '68).

Amphetamines potentiate LSD effects. Nicotinic acid (though not nicotinamide), glutamic acid, and succinic acid counteract LSD effects and can be used intravenously when there is some contraindication to the use of phenothiazines in therapy (Clin Pharm 6:183 '65).

Morning glory seeds contain several ergot alkalloids including lysergic acid amide. In addition to LSD-type intoxication, morning glory seed gnashing may be followed by signs and symptoms of ergotism.

HERXHEIMER REACTION

The Herxheimer reaction is, classically, fever and the intensification or recurrence of syphilitic symptoms within several hours of the initiation of antisyphilitic therapy. A similar reaction is precipitated by specific therapy of some other infectious disease (in all instances it is prevented by relatively small doses of steroids):

1. relapsing fever (indeed, therapy should not be initiated during a febrile spike since the resultant Herxheimer reaction may prove fatal)

2. brucellosis (about 20% of cases treated with tetracycline)

3. typhoid (the Herxheimer reaction in this case is manifested by circulatory collapse without increase in fever)

INSULIN

Topical insulin appears to be a valuable adjunct to the treatment of gangrenous diabetic skin ulcers (Lancet 1:1199 '68).

A common, but often neglected cause, of poor out-patient diabetes control is repeated, faulty insulin injection technique, leading to local incarceration with irregular release into the circulation.

Clues indicative of the Somogyi effect (insulin induced rebound hyperglycemia) are (Am J Med 47:891 '69):

1. asymptomatic periods with urine tests negative for glucose and ketones followed relatively shortly by marked glycosuria and ketonuria

2. wide fluctuations in blood glucose level, frequently unrelated to food intake

3. sharp glycosuric fluctuations from 0 to 4+ without gradual transition

4. glycosuria during much of the day with hypoglycemic symptoms during the night or early morning hours

When insulin requirements change markedly in the post-partum period, Sheehan's syndrome is to be ruled out.

IODINE

Periorbital edema is typical of acute iodide hypersensitivity reactions, which are not uncommon in persons started on iodide containing expectorants. With chronic iodide reactions, patients complain of distasteful mouth sensations, GI changes (especially diarrhea), and increased salivation. They are noted to have submandibular or parotid gland enlargement and an acneform facial eruption.

Contrary to popular belief, tincture of iodine is not particularly toxic by mouth (and is inactivated by food in the stomach). Brown staining of the gums and buccal mucosa suggests iodine has been ingested. When starch is present in the stomach, vomitus will be blue.

LIDOCAINE

Lidocaine does not share antigenicity with procaine amide and can be used safely in patients with prior sensitivity to that agent.

When given intravenously, continuous infusion (25 to 200 mg/hr) is much less likely to cause circulatory depression (particularly cardiogenic shock) than bolus injection, particularly post-infarction. Indications for use are:

1. more than 10 unifocal PVC's per minute
2. more than 5 multifocal PVC's per minute
3. ventricular tachycardia without hypotension
4. recurrent ventricular tachycardia after countershock
5. coupled or tripled PVC's

When infusion of lidocaine causes an apparent increase in the frequency of ectopic beats (and blood pressure is not decreased significantly), it is likely that the beats are of atrial origin with aberrant conduction.

Shakespeare: (Winter's Tale I:2)

I have tremor cordis on me, my heart dances

Shakespeare: (Much Ado III:4)

Get you some of this distilled carduus Benedictus and lay it to your heart; it is the only thing for a qualm.

ORAL CONTRACEPTIVE AGENTS

Use of oral contraceptive agents appears to raise blood pressure minimally in certain individuals. When hypertension is present prior to pill use, patients must be followed carefully for the development of malignant hypertension. Changes which develop are reversed by discontinuation of the medication.

Oral contraceptives should not be used in women with cardiac or vascular disease or conditions in which blood coagulation is facilitated (such as, for example, ulcerative colitis). These agents have a vitamin K-like activity which results in coumadin resistance and acceleration of the pro-time. Hypercoagulability is generally not of functional importance, may depend upon a genetic predisposition, and may also be related to the blood lipid changes induced by these agents. Increased pro-time persists 1-2 weeks after discontinuation of the agent.

Prolonged anovulation may be a sequel to oral contraceptive use. These agents should be given with care to nulliparous women with prior histories of prolonged amenorrhea.

PENICILLINS

Coombs' positive hemolytic anemia (with or without eosinophilia) can occur with large doses of parenteral penicillin. There is no hemoglobinemia. The mechanism appears to be accelerated splenic destruction.

The liklihood of an allergic reaction is the same for all of the penicillins. In many instances, the reaction is to a breakdown product rather than the primary structure, which may explain penicillin-cephalothin cross-sensitivities. Penicillinase is ineffective as an adjunct in therapy of sensitivity reactions. With the vascular collapse of an anaphylactoid reaction, it may be difficult to administer epinephrine via a forearm or jugular vein; in this situation, injection into the sublingual plexus is lifesaving.

The degree of serum binding of a penicillin must be considered when predicting in vivo activity from the results of in vitro testing. When serum binding is high, specific activity is decreased but drug half life is prolonged and the agent IS LESS ACTIVE IN SERUM THAN IN BROTH). Oxa-, clox-, diclox-, and naf-cillin have a high degree of binding, cephalothin, methicillin, and penicillin a low degree.

Methicillin is inactivated in acid solutions. When this agent cannot be given rapidly per volutrole but must be added directly to the IV bottle, 20 meq/l bicarbonate should also be added.

SALICYLATES

Aspirin increases bleeding time and is, therefore, contraindicated in individuals with bleeding disorders. Since the bleeding time prolongation is slight with 'regular' doses of aspirin, this drug can be used safely in patients with normal coagulation mechanisms post-operatively and after childbirth.

Excessive salicylate intake is indicated in children when petechiae are produced by alcohol skin preparation and other mild forms of skin trauma. In adults, irregular, 'blot shaped' subungual hemorrhages may occur with salicylism.

117

GI mucosal irritation with chronic salicylate use leads to a slight degree of occult blood loss in, perhaps, most patients on such regimens. This factor may be of some importance during childhood, pregnancy, and when patients are already anemic from other causes.

Aspirin absorption from the stomach is speeded, and therefore, 'relief-time' is decreased, when aspirin is taken orally with hot water.

The signs and symptoms of salicylate poisoning are relative to the dose per body weight (or body surface area) which is ingested (apropos, aspirin is the second most often used drug in suicidal attempts).

With mild aspirin overdose there is:
1. increased rate and depth of respiration, leading initially to respiratory alkalosis and increasing the rate of insensible water loss
2. central nausea and vomiting
3. drowsiness and confusion
4. sweating, thirst
5. tinnitus, decreased auditory and visual acuity

With more severe intoxication there is, in addition, epinephrine release, impairment of carbohydrate metabolism, and uncoupling of oxydative phosphorylation leading to accumulation of lactic, pyruvic, and acetoacetic acids, all of which may suggest diabetic ketoacidosis initially.

With very severe intoxication, respiration is depressed (CO_2 retention occurs) and mental status is severely deranged.

Alkalinization of the urine enhances salicylate excretion. Forced diuresis including sodium bicarbonate and potassium supplements in the fluid regimen appears to be the treatment of choice of salicylate intoxication.

STEROIDS

Insomnia and disturbing dreams precede the development of overt psychosis in patients given high doses of steroids.

Slocumb et al (Ann Int Med 46:86 '57):

The characteristic feature of chronic hypercortisonism is the appearance of cyclic swings of mood and symptoms. Increased fatigability, muscular and joint aching, and emotional instability alternate with periods of stimulation associated with a "restless drive" and relatively little aching. The relative predominance of either part of the cycle varies from patient to patient..... When depressed, the patient may be unable to concentrate normally; cooperation as well as activity is restricted. Spontaneous crying spells may occur. In general, the patient feels better with rest and worse with fatigue. Activity such as shopping and cooking may be interrupted by quickly developing severe physical and mental fatigue.

A complication of long term steroid therapy which is often neglected in the follow up examination is the development of a (posterior subcapsular) cataract. Chronic topical steroid application to the eye is more likely to cause a cataract than is systemic administration.

Chronic hypervitaminosis A and D may both follow overzealous vitamin use. The features of hypervitaminosis D follow calcium mobilization from bone and are chiefly those of hypercalcemia (see "CALCIUM").

With chronic hypervitaminosis A there is:
1. irritability, fatigue, apathy

2. coarse, dry skin and hair, dry fissured lips (eventually there is loss of scalp and body hair and eyebrows), pruritis

3. waxing and waning bone aching (in children, bone X-rays reveal irregular periosteal thickening, typically involving the ulnas and clavicles; though not pathognomonic, such X-ray changes are quite suggestive).

CSF production is increased, and in infants, neurological findings consequent to increased intracranial pressure may dominate the presentation. In adults there are rarely neurological signs, although a non-specific recurrent headache may occur.

The dyad of neurological abnormality and irregular periosteal thickening should also suggest (in the infant) Caffey's disease.

Individuals with hypovitaminosis A have long eyelashes and are frequently found to have hyposmia. The initial sign of this deficiency in children is conjunctival dryness.

INDEX

central nervous system, 36, 87

Jacod's syndrome, 28
Jaundice, 57
  chlorpromazine induced, 111
  with renal disease, 80
Jaw:
  See Mouth, 27
Jervell Lange syndrome, 13
Jugular vein distention:
  See Rheumatic heart disease, 81

Kallman's syndrome, 36
Kaiser-Fleisher rings, 12, 95
Kehr's sign, 13
Keratoses (palmar):
  with arsenic toxicity, 110
  with pachydermoperiostosis, 69
Kidney disease, 80
Klinefeiter's syndrome, 36

L forms, 97
Lacrimal gland, 16
Lacrimation:
  increased with roach poisoning, 32
  See Tears
Lactase deficiency, 14
Lactic acidosis, 10
Laryngectomy:
  olfactory changes, 29
Latex fixation, 100
Laurence Moon Biedl syndrome, 36
Laxative abuse, 14
Left axis deviation, 109
  See also Congenital heart disease, 48
Leg, 24
Lens, 25
  See Cataract
Leonine facies, 17
Leprosy, 61
Lesch Nyhan syndrome, 61
Leukemia:
  bone x-ray changes, 107
  monocytic leukemia, 62
Licorice,
  inducing hypertension, 59
Lidocaine, 115
Liver disease, 57
LSD, 114
Lung disease, 77, 78, 79
Lupus erythematosus, 62
  See also Cornea, 12

See also Cytoid bodies, 12
Lymphatic hypoplasia, 24
Lymphocytosis:
  with Addison's disease, 37
Lysosomes, 63

Macrocytosis, 101
Magnesium, 101
  and digitalis toxicity, 112
Malaria, 87
Marfan's syndrome:
  lens dislocation, 25
  mitral prolapse with, 64
Maroteau-Lamy syndrome, 65
Measles, 28
Median neuropathy, 47
Melanoma:
  examination for, 33
Melanosis oculi, 24
Melkersson Rosenthal syndrome, 27
Meningioma:
  proptosis indication, 29
Meningitis:
  cranial bruit with, 33
Metastasis:
  anticoagulant therapy, 112
Migraine, 22
Milk alkali syndrome, 98
Miosis, 31
  in coma, 11, 117
  in Horner's syndrome, 59
Mitral facies, 17
Mitral insufficiency, 20, 63
Mitral prolapse, 63
Mitral stenosis, 81
Monocytic leukemia, 62
Mononucleosis, 64
  epitrochlear nodes with, 26
Monosodium glutamate:
  See Chinese restaurant syndrome, 47
Morning glory seeds, 114
Morquio's syndrome, 65
Mouth, 26
Mucopolysaccharidoses, 65
Mucormycosis, 51
Multiple myeloma, 65
  vertebral involvement vs.
  metastasis, 107
Multiple sclerosis:
  anisocoria with, 32
  facial myokymia with, 18
  internuclear ophthalmoplegia
  indicating, 16
  macrocytosis with, 101

127